The Blissful Brain

The Blissful Brain

Neuroscience and
proof of the power
of meditation

Dr Shanida Nataraja

An Hachette Livre UK Company

First published in Great Britain in 2008 by
Gaia, a division of Octopus Publishing Group Ltd
2–4 Heron Quays, London E14 4JP
www.octopusbooks.co.uk

ISBN 978-1-85675-291-6

A CIP catalogue record for this book is available from the
British Library

Printed and bound in Italy

Printed on Cyclus Offset, a 100 per cent recycled paper

10 9 8 7 6 5 4 3 2 1

All reasonable care has been taken in the preparation of this
book, but the information it contains is not meant to take
the place of medical care under the direct supervision of a
doctor. Before making any changes in your health regime,
always consult a suitably qualified doctor. Any application of
the ideas and information contained in this book is at the
reader's sole discretion and risk.

Dedication

I dedicate this book to my family – my parents, Kim and Shankar, and my brother, Ramesh. Without their love and support, this book would have not been written.

CONTENTS

Acknowledgements

First and foremost, I give my heartfelt thanks to my partner, James Yates, who has been by my side throughout the highs and lows of writing this book. His support, and measured criticism, has kept me on my path and made the book what it is today.

My warmest thanks also go to my mother, Kim, and my old friend, Sifu Dr Mark Green, both of whom contributed to and inspired me to write this book. Their impassioned application of contemplative practice to everyday modern life is an example for us all.

I also extend my gratitude to my agent, Stephanie Ebdon, whose belief in the potential of this book and the importance of the messages it conveys has helped to drive the book's development and eventual publication.

And finally, my thanks go to everyone who has, in some way, contributed to the writing of this book: 'American Dave' for first tweaking my interest in mystical technologies; my teachers and scientific colleagues over the years for igniting my passion for the wonders of science; and my friends on both sides of the Atlantic who have endured my ranting and ravings on this topic for years.

Introduction

'ENLIGHTENMENT AT THE FLICK OF A SWITCH' was the hook he used to grab my attention and entice me to a warehouse in a rundown and frankly dangerous part of Baltimore, Maryland. At the time, I had just started the second year of a two-year post-doctoral position in the Department of Neuroscience at Johns Hopkins University School of Medicine. Working in the laboratory of Dr David Linden, I was attempting to unravel the cellular processes underlying learning and memory in the rat brain. My research, although challenging, seemed so distant from the 'real world', so detached from the human behaviour that I had initially set out to understand better. Day after day was spent with my eye glued to a microscope, inserting tiny glass electrodes into slices of brain tissue, desperately hoping that, even under such unnatural conditions, some of the brain cells would still be alive and healthy and suitable subjects for experimental scrutiny. Somewhere, deep within these slices of brain tissue, supposedly lay the key to understanding how we, as humans, learn new behaviour, how we modify or abolish existing behaviour, and how we store and recall learned behaviour for, in some cases, the length of our lifetime. However, after countless hours in the laboratory, I realized I wasn't going to find this key by examining the behaviour of single brain cells. The answers, at least to my own questions, would not be provided by this dissected approach to the study of the human brain and how we think and act.

He was a twenty-something guy from New Jersey. Diagnosed with attention deficit hyperactivity disorder (ADHD) at a young age, he was, like many others with this condition, incredibly bright, if a little unfocused. He had dedicated much of his adult life so far to finding a way of dealing with his condition that didn't involve dosing himself daily with Ritalin®. In this quest, he had stumbled across technology that he claimed would not only allow him to combat his ADHD, but would also allow anyone, in time, to enhance the performance of their brain, and thus their mind. On that first night – one of those hot and humid nights in Baltimore that makes everyone hit the streets as sleep is impossible – he introduced me to the galvanic skin response (GSR) meter; one of a number of 'toys' he had found on this quest. We sat on an old sofa in the middle of a huge, largely empty, warehouse space, as he attached an electrode to one of my fingers on my left hand and fired up his laptop.

In the hours that followed, I learned that the GSR meter detects the level of arousal in the brain and can therefore be used as an indirect measure of one's stress response to different situations or trains of thought. For the first time, I was able to see an external visualization of what my state of mind was doing to my body, watching how different breathing techniques could lower the level of arousal, and thus release tension, and how certain thought patterns could raise it and thus create tension. The GSR acted like a mirror for my state of mind. I gained conscious awareness of the constant state of arousal, or stress, in my body, and could test the effect that different relaxation techniques had on that level of stress and indeed my clarity of mind. Here was an instrument, widely available over the internet, that the general public could use to gain a greater awareness of their behaviour, a greater awareness of their internal state of mind.

And this 'toy' was merely the tip of the iceberg. Although it soon became clear that technology had not provided us with a fast-track route to enlightenment, this encounter prompted me to start a quest of my own. The last 30 years have seen an enormous leap in our understanding of the human brain and behaviour. A wealth of technology, of increasing complexity and precision, has allowed us to probe ever deeper into the workings of our brain. We have defined the cellular processes underlying learning and memory, we have assigned individual aspects of behaviour to different regions of the brain, and we have sketched out some of the dynamic processes that give rise to the wealth of human behaviour. The 1990s were designated the 'Decade of the Brain', and this pivotal period in neuroscience research hosted a series of major scientific breakthroughs that revealed the neural processes underlying complex human cognitive functions, such as learning and memory, the neural basis of altered/higher states of consciousness, and potential ways of optimizing the performance of the human brain. Slowly, we have gained a fuzzy picture of how the human brain works, and how it acts to define, execute and coordinate our complex behaviour.

The implications of this research for everyday human behaviour, however, have been less well defined. I therefore set out to discover what these scientific findings reveal about human behaviour, and specifically what clues they provide about how we could possibly improve the performance of our brains and so express more of our full potential as human beings. This journey led to meditation.

Throughout history, we have sought to expand our awareness using contemplative techniques such as meditation, rituals such as the dance of the whirling dervishes, and by using mind-expanding drugs. Recent research has revealed processes in the human brain, and in the human mind, that appear to underlie these routes to enhanced awareness. Similar processes occur in the brain when a person meditates and when they engage in any ritualized behaviour, whether it be the repeating of a prayer or chanting at a football match. In many cases, it has been possible to observe, in real time, the brain processes involved in generating these altered states of consciousness, including mapping those that arise through meditation.

Countless contemplative traditions claim to teach methods of expanding awareness, gaining self-knowledge and realizing human potential. With advances in technology, these traditions have now come under scientific scrutiny, and a growing body of evidence points towards the benefit of regular meditation, and the role that it plays in enhancing human performance and maintaining good health and wellbeing. Furthermore, as our scientific understanding of meditation has deepened, we have also found possible ways of facilitating existing contemplative methods and avoiding their inherent pitfalls. The combination of techniques developed and fine-tuned over many centuries with recent technological advances offers a viable strategy for anyone who wants to gain self-knowledge and a greater awareness and control over their behaviour.

This book is a scientific guide to meditation, which reveals that practices such as meditation, tai chi and yoga not only play a crucial role in optimizing our performance and wellbeing, but may also significantly reduce stress and produce substantial health benefits. Stress, and stress-related diseases, place an increasing burden on our Western healthcare systems, and there is a clear need for a shift in our approach to health, both as individuals and as a society. The widespread acceptance of meditative practices as a crucial component of effective patient management has, until recently, been hindered by a lack of knowledge about the effects of meditation on the brain and on measurable health outcomes. However, medical evidence suggests that meditation has measurable effects on stress levels, as well as reducing heart rate, blood pressure and the risk of cardiovascular disease, and in boosting the immune function, melatonin levels and psychological wellbeing. Not only does this warrant increasing the provision of meditation training as part of patient care, but it also supports the inclusion of meditation in our everyday lives, both as a preventative health strategy and as a means by which to realize our brain's potential and enhance our wellbeing.

Meditation

What is it and why do we need it?

MODERN LIFE IS STRESSFUL. Not only does this have a negative impact on our sense of wellbeing, but it also places an increasing burden on our healthcare systems. More and more people are succumbing to stress-related diseases and healthcare-related costs are climbing at an alarming rate. There is also an increasing body of evidence indicating that the conventional Western approach to healthcare is inherently flawed and incapable of meeting current demand. There is therefore a clear need for a more holistic, patient-centred approach to disease management, and this chapter presents evidence that the necessary shift is already in motion.

We have seen a growing interest in practices such as meditation, tai chi, chi kung (also referred to as qigong or chi gung) and yoga. These disciplines appear to go some way towards offering an antidote for the stressful environment of modern life, and are seen by many to be a useful adjunct to conventional patient care in a wide range of medical conditions.

What is meditation?

Although virtually everyone has heard of meditation, tai chi and so on, we may actually have quite a confused idea of what they actually are. Although these practices are Eastern in origin, they have much in common with Western traditions of prayer and ritual.

Meditation in the Buddhist tradition involves a process of intense concentration and attention to quiet the conscious mind. There are countless variations on this process, and most can only really be understood within the context of the particular culture in which they are practised. Dzogchen meditation as practised in Tibet, for example, differs considerably from the method of the same name that has gained popularity in the West in recent years. However, all meditative methods have common features, which provide us with a key to an appropriate definition of meditation. In order to be considered a meditative discipline, a procedure must meet five essential criteria:

- it must involve a specific technique that is both clearly defined and taught
- it must involve, at some stage, progressive muscle relaxation
- it must involve, at some stage, a reduction in logical processing
- it must be self-induced
- it must involve a tool, referred to as an anchor, that allows effective focus of the mind.

There is an abundance of different meditative methods, but chiefly they fall into one of two categories: top-down and bottom-up. Top-down methods confer expanded awareness through relaxation of both the mind and body, and include such methods as meditation and contemplative prayer, whereas bottom-up techniques elicit the same expanded awareness through prolonged activity arousal of both the mind and body, and include dancing, chanting, sleep deprivation and some forms of vigorous yoga.

Over the last 40 years, the bulk of the research into contemplative practice has centred on two meditative techniques: transcendental meditation (TM) and mindfulness-based stress reduction or MBSR (see below). Through this research, it has become clear that the key to all of these meditative practices is attention.

Mindfulness-based stress reduction (MBSR)
MBSR is a meditation technique first proposed by Jon Kabat-Zinn and his colleagues in 1979. It has been investigated as a possible therapeutic option in patients suffering from physical, psychosomatic and

psychiatric disorders. Although this technique is taught independently of any religious or esoteric tradition, it is deeply rooted in the contemplative spiritual traditions, in particular the Buddhist mindfulness meditation. Jon Kabat-Zinn's own definition of mindfulness is: 'awareness that arises through paying attention on purpose, in the present moment, and non-judgementally to the unfolding of the experience moment by moment.'[1]

Its techniques, as an example of the practicalities of meditation, are described more fully in Chapter 7 (see page 194).

Variations on a theme

We may tend to think of practices such as tai chi, chi kung and yoga as primarily physical disciplines, but they actually share many of the key features of meditation, including the focus on attention, and bring about similar benefits. Let's take a quick look at some of these disciplines, beginning with the one that is probably the best known in the West, yoga.

Yoga

In a recent US survey of 2,055 adults, 7.5 per cent reported having used yoga at least once in their lifetime and 3.8–5.1 per cent reported having used it in the previous 12 months.[2] But the much-practised postures of yoga are only one aspect of the yogic philosophy, which encompasses a whole way of life. Ancient texts identify eight elements of yogic practice:

1 **Yama**: social behaviour; how to treat others and the world as a whole
2 **Niyama**: inner discipline and responsibility; how to treat ourselves

3 **Asana:** practice of postures; how to prepare the body for meditation

4 **Pranayama:** control of the breath; how to control the flow of energy, or *prana* through the body

5 **Pratyahara:** withdrawal from the senses; how to withdraw and direct attention inwards during meditation

6 **Dharana:** concentration; how to train the mind to one-pointed focus

7 **Dhyana:** non-goal-orientated meditation; how to meditate

8 **Samadhi:** complete union or absolute bliss; the ultimate 'goal' of the yogic path

The word 'yoga' literally means 'union,' and is derived from the Sanskrit word *yuj*, meaning 'to unite'. Yoga is seen to reunite mind and body, stillness and movement, masculine and feminine.

Types of yoga differ considerably with respect to what they involve, both in terms of lifestyle and daily practice. In the West, we mainly associate yoga with the physical postures, or asanas, which are known as hatha yoga, and when you attend any type of yoga class, you are actually practising hatha yoga. Within hatha yoga there are several different yogic schools, of which astanga vinyasa, iyengar and kundalini yoga are the best known throughout the world today.

Iyengar yoga is perfect for a beginner or someone with limited flexibility. It comprises a series of postures, controlled breathing and meditation, which, over time, brings greater flexibility and relaxation. Props such as chairs, straps and pillows can be used to compensate for any initial lack in flexibility. Astanga yoga is designed to build stamina and strength, so involves moving rapidly from one posture to the next; it places limited emphasis on the meditative component. Kundalini yoga is designed to give greater control of body energy. The word *kundalini* literally means 'the curl of the lock of hair of the beloved', or coiling – in the classical literature of hatha yoga, kundalini is described as a coiled serpent at the base of the spine. The image conveys the sense of the untapped potential energy within the human being.

Chi kung

In the West chi kung is primarily understood as a bodywork approach following a system of gently flowing movements; however, it has ancient origins as a healing therapy that can be applied spiritually, medically and even as a martial art.

The earliest documented evidence of chi kung is found in the Yellow Emperor's *Classic of Internal Medicine*, written during the Han Dynasty, 206 BC–AD 220. The era also bequeathed another significant document to the history of chi kung, the *Dao Yin Tu*. (Authorship is attributed to Hua Tuo, a popular figure in the history of traditional Chinese medicine who has been described as the patriarch of Chinese medicine and the first famous Chinese surgeon.) The *Dao Yin Tu* includes a diagram of 44 human figures performing exercises that emulate the movement of animals, including the wolf, monkey, bear, crane, hawk and vulture. This imitation of animal movements has been made famous in modern times by the stage performances of the Shaolin monks and provides us with a clue to the ancient origins of chi kung: the ritualistic practices of Siberian and Mongolian shamans. Chi kung is integral to traditional Chinese medicine, and its movements are thought to stimulate the acupuncture meridians (see page 39).

Today, chi kung has three major schools: spiritual, medical and martial. Despite this diversity, the basic goals remain the same: building strength, maintaining good health, curing disease, prolonging life and cultivating spiritual awareness.

Some methods are physically very simple, others consist of sequences of flowing movements that are integrated with natural, relaxed and rhythmic breathing and combined with a contemplative state of mind. In this form the method has both a physical and spiritual component and, like yoga, its overall benefits are similar to meditation.

Chi kung's physical component (body movements) and spiritual component (meditation and intention) are separate elements of the practice, and it is interesting to see how the development of chi kung practitioners is clearly influenced by their intention, why they are practising. For example, if you practise chi kung for healing, you will not develop the internal power needed for martial arts. On the other hand, if your intention is fully focused on gaining martial power, you may not progress spiritually.

Tai chi ch'uan

Tai chi uses the basic system of chi kung but superimposes it over the framework of a martial art – the movements of tai chi are both combat techniques and healing chi kung. In the West, however, tai chi is commonly taught in terms of its contemplative and healing aspects and in this way it is seen as 'meditation in motion'.

The first significant blending of the principles of traditional Chinese medicine and martial arts was seen in the teaching of Chang San Feng in the 14th century. Chang, who was a master of both acupuncture and martial arts, spent many years living as a Taoist hermit crafting his own martial system, which he named Wu Dang Mountain Boxing (after the mountains in which he lived).

Yang Lu Chan (1799–1872) became a great master of the Wu Dang system and in later life created his own style of martial art in which, like Wu Dang Boxing, each movement both promoted the flow of chi and was a martial arts technique that targeted the vulnerable acupuncture points of the opponent's body. The effectiveness of Yang's art was so profound that later generations named it tai chi ch'uan or 'supreme ultimate boxing'. Yang's grandson, Yang Cheng Fu, simplified his grandfather's art with an emphasis on the healing and contemplative aspects. Being much easier to learn, this version of tai chi became immensely popular, growing into a global movement that now boasts over 100 million practitioners.

There are now many different styles of tai chi. Generally speaking, however, all of them comprise a series of individual postures seamlessly joined by transitional movements, performed in a slow and graceful manner. Like chi kung, tai chi also involves a psychological component and requires a contemplative state of mind. Practitioners concentrate on a number of different elements: the relaxation of the body; the shifting of the weight from one leg to the other as if wading through water; the movement of the limbs and so on. In much the same way as a meditative mantra or chant can anchor the attention, focusing on these specific movements also helps to bring the mind to focused attention, blocking out extraneous thoughts. And, as with chi kung, movement is synchronized and defined by breathing, acting as another anchor for meditative attention, and the intention of the practitioner determines how they progress over time.

Meditation techniques evolved in the West

Although we often associate meditative practices with Eastern philosophy, contemplation and centering prayer have been a prominent part of daily Christian life for many centuries, and contemplative orders of nuns and monks are still scattered across the globe. Throughout the scriptures there are references to the essentials of the contemplative path: silence, interiority and the use of few words in prayer. As Christianity became the religion of the Roman Empire under Emperor Constantine in the fourth century AD, the 'literalist' strand of Christianity, based on the Gospels, flourished. However, many Christians retired to the silence and solitude of the desert in search of the opportunity to experience an authentic spiritual life. Contemplative prayer and meditation flourished in these remote locations and have re-emerged at different times and in different forms since.

Christian meditation

The desert tradition stressed the importance of purifying the desires – fighting the 'demons' and freeing the mind from the drives of the 'ego'. The way to this purification was detachment, not only from thoughts of self but in some cases even thoughts, words and images about God. A formula, or prayer phrase (i.e. a mantra) would be used to silence the mind, leading to pure contemplation without conscious awareness.

The teaching of the early desert Christians was written down and brought to the West by John Cassian. However, owing to the orthodox nature of the Church at the time, the practice of mantra meditation was largely lost by the sixth century AD, surfacing only occasionally in the writings of Christian mystics. In the 20th century a Benedictine monk, John Main, rediscovered mantra meditation in the writings of John Cassian and the practice of mantra meditation has once again become a viable and widespread component of Christian spiritual life.

Centering prayer

The centering prayer of modern Christianity has its origins in the teaching of three 20th-century Trappist monks. Their teaching, like that of John Main, is largely based on that of the early Christian mystics, and centering prayer superficially follows a similar format to mantra meditation. A chosen sacred word is repeated to bring about a quieting of thoughts and feelings in order to experience the presence of God. However, unlike mantra meditation, in which the mantra is used to focus the attention, the sacred word is used as an expression of intention and consent to engage in contemplative prayer. Despite this conceptual difference, the mantra or sacred word acts as an anchor in both cases. By drawing attention back to the sacred word, a person praying does not become absorbed or distracted by thoughts that inevitably arise during a period of contemplation. At times, the sacred word may become vague or completely disappear: at this point the practitioner 'rests' in the silence.

Lectio divina

The reading of scripture reflectively, as a source of spiritual insight and growth, is an integral part of Christian tradition and was formalized in the early sixth century AD by St Benedict. Initially, *lectio divina* was practised in a group, but as literacy and solitary reading became more common, it became something that could be practised in private.

Lectio divina is comprised of four main parts: *lectio* (the passage is read several times); *meditatio* (the text is reflected upon and applied to everyday life); *oratio* (the formulation of a prayer or repeated phrase); *contemplatio* (silent, wordless contemplation). The text is not treated as a vehicle of information to be analysed, but the emphasis is on engaging with the text in a way that may potentially change awareness of self, the world and the 'ultimate reality'.

Meditation in its many forms continues to be used in the East to promote spiritual, mental and physical benefits. Today, in the West, we seem increasingly in need of the benefits such practices confer.

The stresses of modern life

In his acclaimed documentary, *Bowling for Columbine*, Michael Moore interviews the California sociologist Professor Barry Glassner. Glassner portrays the United States as a country controlled by the media and accordingly gripped by fear. He reports that crime rates in most of the major US cities fell significantly in the 1990s; however, as media reporting of crime increased dramatically over the same time period, the

American public remained fearful. Threats of violent crime have been joined by threats of terrorism, infectious diseases and environmental catastrophes, among others.

This climate of fear is not restricted to the USA; similar trends can be seen across the world. Every day our news is dominated by warnings of potential threats to our survival. The increasing level of fear in the general populous creates an extremely stressful environment. We are constantly primed to respond to the next potential threat, whether that be running away from a would-be attacker or responding to a bomb alert in a train station. Added to this, pressures resulting from competition in the workplace, economic hardship, social unrest and conflict, and poor working and living conditions all create stress, as do the increasing demands on individuals to juggle all of the many different elements that constitute our modern-day lives.

The world has always been a stressful place. Our primitive ancestors had to deal with continuous threats to their survival. Over time, their brains evolved to handle these life-threatening situations better, and their learning is encoded in the circuitry of the more primitive regions of our own brains. However, in our modern world, the level of stress appears to be escalating.

The industrial and technological growth in the last couple of hundred years has undoubtedly improved life expectancy in the West; however, this has been at the expense of the world's poorer nations, our natural environment and our health. With the advent of advanced telecommunications, and relatively easy global travel, people in the West have become more aware of the non-sustainable nature of their lifestyles. We are bombarded by information about horrific working conditions in the sweatshops in the Far East now producing our designer wear, about people across the globe who have lost their livelihoods as a result of climate change, deforestation, agricultural practices, species extinction, desertification and urban growth, and about the worldwide consequences of greed, corruption and the never-ending search for the 'pot of gold at the end of the rainbow'. For decades, ecologists have warned us of the non-sustainable nature of our global practices, pointing to evidence of climate changes, decreases in biodiversity and an increasing strain placed on our planet's resources. Now, at the beginning of the 21st century, we are faced with glaringly obvious examples of an ecosystem in crisis. Whereas our ancestors were escaping from sabre-toothed tigers and hunting wild bison, the stress in our lives largely stems from our non-sustainable approach to all of our endeavours. It is not good enough to party, we must 'party hard'; the 9 to 5 working day has been lost in many professions, giving rise to a generation of workaholics; we eat and drink too much, spend too much on items we do not need and devote far too much time doing rather than being.

The costs of stress

Unsurprisingly, the increasing level of stress in our world has been accompanied by an explosion in the incidence of stress-related diseases, such as heart disease, high blood pressure, depression and anxiety. Stress also increases the risk of these diseases by increasing risk-prone behaviour, such as excessive alcohol consumption, smoking, overeating, physical inactivity and the use of illicit drugs. We are faced with a number of worrying health epidemics, many of which can be linked with stress and our resulting unhealthy lifestyles.

Cardiovascular disease remains the most common cause of death worldwide, despite an increased emphasis in recent years on the importance of modifying risk factors, such as smoking, obesity and a sedentary lifestyle. Similarly, depression and mental health disorders in general are becoming increasingly prevalent. In the UK it is estimated that about 15 per cent of the population are clinically depressed; in reality, this figure is thought to be much higher, as it is known that a significant proportion of people with depression do not seek out professional help. The World Health Organization (WHO) predicts that depression will be the second leading cause of death worldwide by 2020.

Psychosomatic disorders – those that involve physical symptoms, but have an emotional or psychological origin – are also commonplace, particularly in the West. Psoriasis, eczema, stomach ulcers and irritable bowel syndrome (IBS) have all been shown to be triggered and exacerbated by factors such as stress and anxiety. The term 'psychosomatic' is often used in a negative sense, to imply that the perceived symptoms are in some way created by the individual, and should therefore be ignored. However, it is now widely accepted that psychosomatic conditions should receive the same attention as those conditions that have an identifiable physical cause. IBS was, until recently, not recognized as a physical disorder, despite causing considerable patient discomfort. It is now thought that 10–15 per cent of the general population suffer from IBS. As another example, the number of new cases of asthma has increased dramatically in the last 25 years. While this is probably the result of increased environmental pollution, excessive hygiene and poor dietary habits, an increased severity and frequency of asthmatic attacks is linked with an increasing level of stress.

The financial costs associated with stress, specifically stress occurring in the workplace, have also been estimated. About 40 million workers in the European Union are affected by work-related stress. The most common causes are long hours, shift work, job insecurity and bullying, and the costs of working days lost as result of this stress amount to more than €20 billion annually. Work-related stress does not merely affect high-flying 'city types'; all of us experience stress resulting from our workplace or career at some point in our lives. The cost of this stress extends beyond its impact on work productivity; it wreaks havoc on our personal relationships, on our physical and mental health, and on our personal finances as we become increasingly dependent on the many quick stress-relievers our society has to offer: fast food, alcohol, smoking, prescription and illicit drugs, shopping and sex.

A shift in lifestyle

It is perhaps not surprising, then, that in these stressful and confusing times an increasing number of people are questioning their lifestyle and career choices. Until the 1970s, the large majority of people settled on a career and then climbed the career ladder, usually in the same company, until retirement. These days, job changes every couple of years are the norm, not the exception, and many people switch career several times during their working lifetimes.

As an increasing number of people have achieved, through successful careers or business ventures, high levels of material security, there is a growing emphasis on physical, psychological and emotional wellbeing over material luxuries. Wealth and security appear not to guarantee happiness and people are turning to other sources for a sense of wellbeing and personal fulfilment. These trends are reflected in an increase in the number of people opting for meaningful careers over well-paid careers, the growth of the health and fitness industry, and a shift in our consumer habits towards environmentally friendly, organic, fair-traded and renewable produce. The overwhelming evidence pointing towards the detrimental effects of our stress-filled approach to life in the West questions the continued emphasis placed on financial security and material goods over our physical, psychological, emotional and spiritual health. It is well known that meditation, yoga and similar practices can aid relaxation and reduce stress. But new research is revealing that stress reduction may be even more vital for our total health and wellbeing than previously thought. There is now compelling evidence of the toll that long-term stress takes on our health, and the importance of both mind and body in healthcare.

The relationship between body and mind

The 17th-century French philosopher, René Descartes, is responsible for the age-old division of mind and matter in the West. Descartes divided nature into the realm of the mind, *res cognitans*, and the realm of matter, *res extensa*. His famous statement *'Cogito ergo sum'*, 'I think, therefore I am', defined Western thought for centuries, and human beings were labelled as isolated minds, separated from God, from their physical bodies, from each other and from their environment.

Most modern thinkers now reject the view that mind and matter are independent entities. There is a growing body of evidence that suggests that mind influences matter, and vice versa. One of the first scientific explorations of the mind–body relationship was conducted in patients with chronic depression. These patients were found to have much higher mortality than the general population, largely owing to cardiovascular disease. This observation can be partially explained by the fact that patients with mental health disorders frequently display at-risk behaviour (such as smoking, the consumption of illicit substances and physical inactivity) that confers a greater risk of developing further medical or mental conditions. It can also be partially explained by the known health risks associated with many of the medications used to treat patients with mental health disorders.

The psychological impact of physical disease has also been explored in depth. It is now well known that patients with chronic or severe medical conditions often need psychological support in addition to physical medical treatment to tackle the depression and anxiety caused by their illness. There is also evidence that patients with a more positive outlook (and thus better coping mechanisms) have better long-term disease and treatment outcomes than those with less developed coping mechanisms.

The clear link between a patient's psychological state of mind and physical health warrants the active inclusion of psychological or spiritual support in patients' care plans in addition to the existing, largely physical treatments. Chapter 6 looks in detail at the role meditation can play in restoring and maintaining our physical and mental health. In particular it looks at conditions that are known to be exacerbated by stress, ranging from high blood pressure to depressed immune function.

The effect of our hormones: the neuroendocrine system

The main physical link between our mind, and therefore our psychological state, and our body, is believed to be the neuroendocrine system. This is a network of nerve cells that produce and secrete hormones into the bloodstream. The release of different hormones is triggered by activity in the brain, and so changes in brain activity can drive changes in the body's functioning.

Stress can trigger long-term abnormalities in the neuroendocrine system. A stressed individual is in a state of permanent arousal, ready to 'fight or flee' when necessary. As a result, levels of the stress hormone, cortisol, are usually higher than normal. This can lead to memory loss – when stressed we are more likely to forget something of importance, such as a spouse's birthday or an important deadline. Sustained or extreme stress has also been linked to chronic depression and anxiety. This is perhaps best illustrated by post-traumatic stress disorder (PTSD). People who have had a traumatic experience, such as being physically attacked or mugged, often report feeling emotionally detached or numb after the event. Some also report feeling irritable and having difficulty in sleeping and concentrating. Sufferers are also often found to be clinically depressed and anxious, as well as being dependent on alcohol, prescription and illicit drugs, and smoking.

The link between stress and mood is thought to be the result of stress affecting the amount of serotonin in the brain.

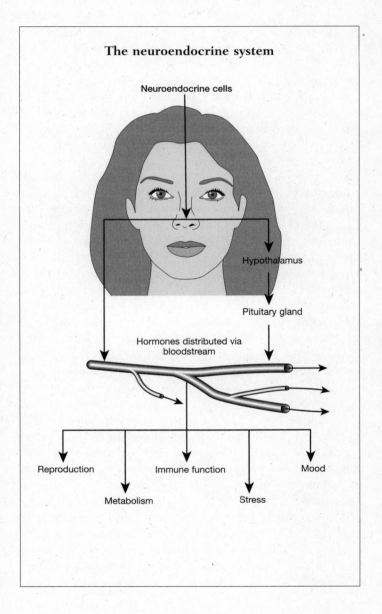

The neuroendocrine system

Neuroendocrine cells

Hypothalamus

Pituitary gland

Hormones distributed via
bloodstream

Reproduction

Metabolism

Immune function

Stress

Mood

Serotonin is a chemical that influences an individual's emotional state: a happy state is associated with increased levels of serotonin and an unhappy state is associated with decreased levels. If you are stressed, the levels of serotonin in your brain can become significantly reduced, leading to depression. Unsurprisingly, the increased level of stress in our world has been accompanied by an increasing prevalence of depression and anxiety. And as we, as a society, become unhappier, we also tend to seek solace in activities and substances that provide a temporary burst of serotonin, such as food, alcohol and recreational drugs.

Stress can also lead to a compromised immune system. The activation of the neuroendocrine system releases hormones that reduce immune function. People under stress often feel 'under the weather', are more likely to pick up the 'flu virus going round the office, and are at greater risk of a number of other opportunistic infections. In a study carried out in 2002 by Sheldon Cohen and colleagues, for example, stress was seen to depress the immune system, leaving people more susceptible to upper respiratory illness. In the study, subjects were asked to defend themselves against an alleged crime, such as shoplifting or a road traffic violation. During this period of stress, some had raised cortisol levels, resulting in a depressed immune system. Others, who exhibited only small rises in cortisol, showed little change in their immune function.[3] These results suggest that people can respond differently to stress: some people have a large physiological response to stress, and its long-term consequences are more severe; other people have a small physiological response and experience only mild consequences.

If stress plays a very active role in our health, affecting our endocrine and immune systems, our blood pressure and our cardiovascular health, is our current approach to healthcare doing enough to address the problem of stress? The answer seems to be no.

The traditional Western approach to health

The Greek philosopher Hippocrates is considered to be the founding father of Western medicine. In the fourth and fifth centuries BC, he taught that medicine was a discipline based on the careful observation and documentation of the patient. Today, this systematic approach to patient assessment still forms the basis of the clinical examination, during which healthcare professionals methodically screen for any signs or symptoms of disease.

Although Hippocrates based his medical knowledge on his observations of human nature in health and disease, the physicians who followed in his footsteps obtained insights through anatomical dissections. Early medics systematically mapped the human body, taking it apart in the same way a mechanic methodically disassembles an engine and labels the component parts. The human body therefore came to be described in terms of the function of major organs and blood and nerve supplies, and the relationship between them.

From the 18th century onwards, medicine was transformed, initially by the widespread use of the microscope, and then by a plethora of different laboratory tests. These advances in technology allowed medical researchers to delve even further into the structure and function of the human body, both in health and in disease, which in turn offered much needed clinical solutions for those afflicted with disease.

Western medicine is therefore founded on a tradition of dissecting the human body into its constituent parts and targeting malfunctioning components for treatment. For some patients, the treatment of symptoms can lead to effective disease management, but since this approach does not target the root or cause of the disease, many patients are required to endure long-term drug treatment and the inherent associated risks. For others, pharmacological treatment only provides temporary or incomplete relief from symptoms, and can result in undesirable and debilitating side effects.

This symptom-focused approach is now slowly being acknowledged to be limited. Someone with an irregular heartbeat undoubtedly requires immediate treatment to restore the normal heart rhythm, as well as medication to maintain it. In the majority of cases, however, an irregular heart rhythm results from underlying cardiac disease, and so effective treatment should involve a full investigation of the fundamental cause and appropriate treatment or cure of the underlying problem. In practice, this is inconsistently achieved – there are simply not enough resources to offer all patients a comprehensive care plan, and limited resources mean that, in many cases, efforts are purely focused on the alleviation or management of symptoms.

Western medicine also tends to be responsive rather than preventative. People rarely seek medical advice until something is clearly wrong and it is obvious that they are in poor health. This means there is little opportunity to prevent disease before it occurs, and by the time the medical problem is acknowledged, there may be little chance to do anything more than slow the progression of disease and treat its troublesome symptoms. We are therefore stuck in a vicious circle.

The changing face of medicine

Modern Western medicine is dominated by an impressive array of diagnostic and therapeutic technology. New imaging techniques and genetic profiling allow better and earlier diagnosis; refined surgical techniques and more effective drugs can now offer extended survival and, in some cases, complete cure.

As the number of diagnostic and therapeutic tools at our disposal increases, there can often be confusion over which offer the best possible outcome to the patient. Guidelines that specify the 'best practice' approach are, increasingly, evidence-based. In other words, the recommendations are based on evidence stemming from the systematic analysis of the findings of medical studies worldwide. This evidence-based approach ensures that current clinical practice embraces recent advances in the understanding of a disease and its treatment, defining a 'standard of care' for all patients.

The recent advances in medical technology have been accompanied by rapidly escalating costs of treatment. Some current treatments for breast cancer cost more than £30,000 a year for each patient, and in 2002 Americans spent more than $12 billion on prescription medicines to combat high blood pressure. The WHO estimates that total healthcare expenditure in the USA rose from 13.1 to 15.2 per cent of the gross domestic product between 1998 and 2003; the equivalent figures in the UK rose from 6.9 to 8.0 per cent in the same time period.

This increase in expenditure is partially related to the availability of more treatment options for patients and the increased costs of developing these treatments. However, it is also partially related to the rising costs associated with the many stress-related health epidemics that burden our healthcare systems today, such as cardiovascular diseases, diabetes, obesity, depression and anxiety. In 2002 the WHO reported that perinatal conditions, lower respiratory infections and HIV/AIDS placed the greatest burden on global healthcare. However, it predicts that in 2020 the top three, in terms of cost to society, will be HIV/AIDS, unipolar depression and ischaemic heart disease.

The traditional Eastern approach to health
In contrast to the Western approach to health, traditional Eastern healthcare is much more holistic, embracing the importance of both mind and body. Rather than targeting the signs and symptoms of disease, Eastern approaches try to identify the source of the underlying problem causing the symptoms and to consider all aspects of lifestyle as being relevant to a patient's health problem.

In Taoism and Neo-Confucianism, the traditional Chinese philosophies, the concept of ch'i is used to represent the energy that animates the universe. Ch'i is thought to condense and disperse in alternating cycles of negative and positive energy. The dynamic interplay between these complementary forces or energies, referred to as yin and yang, is represented by the tai-chi symbol. Yin is the feminine force, the dark pole, and it represents the passive, receptive force. Yang, on the other hand, is the masculine force, the light pole, and it represents the active, creative force.

Yin and yang can be viewed as being polar opposites, but they have a paradoxical relationship: they are opposing, but they are not antagonistic. The opposites share an implicit identity, and are therefore inseparable. 'Light and dark', 'good and evil' and 'life and death' are all seen as complementary aspects of a greater whole. We cannot describe what 'light' is without a reference to its polar opposite, 'dark'; we cannot understand the concept of 'good' without an understanding of its polar opposite, 'evil'. The presence of a stimulus, for example, is meaningless unless there is an opposing concept of the absence of a stimulus, just as without pauses of silence within music, a musical composition has no rhythm.

Trying to understand ch'i

Eastern traditions visualize ch'i flowing through channels or meridians in the human body (see illustration), and poor health is seen to result from disruptions or blockages in this flow. This is the fundamental principle of a number of medical practices, including tai chi, chi kung, yoga and acupuncture, and it forms the basis of the theory of traditional Chinese medicine (TCM).

TCM differs from traditional Western medicine in that its approach is holistic – it looks at the whole body rather than focusing on a specific part of it. Rather than targeting the signs and symptoms of disease, TCM targets the source that gives rise to these signs and symptoms, usually described in terms of

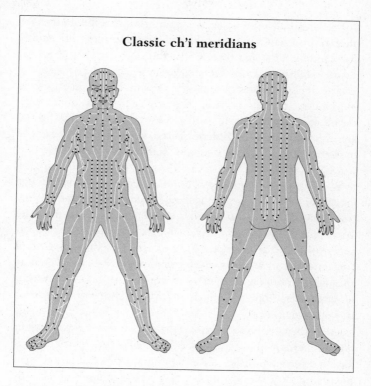

Classic ch'i meridians

specific areas of blocked ch'i. Treatment releases these blockages and allows the body to heal itself at a natural pace.

In TCM, treatment may involve a variety of different approaches, both physical and psychological: a patient may be prescribed a course of acupuncture as well as be instructed to take herbal medicine and practise specified chi kung exercises. The actions of all these complement each other and aim to restore the natural flow of ch'i through the body.

Ch'i is seen to be something that is distinct from the circulatory and nervous systems, but it is undeniably related to both. It is seen as having a complex relationship with emotions; suppressed emotions are seen to be the cause of blockages in energy or ch'i. However, these emotions cannot be equated to ch'i; the constant swirl of emotions actually degrades the ability to focus the mind and therefore to direct ch'i.

Ch'i is also intimately related to consciousness in that it is said to follow the mind. Any action, whether it's a mundane movement such as picking up a pen from a table or a contemplative manoeuvre such as those found in chi kung and tai chi, is always preceded by the intention to make the movement. In the 2001 Richard Kelly film, *Donnie Darko*, a troubled high-school kid suffering from hallucinations imagines that he can see a fluid column of liquid energy extending from his chest, travelling ahead of him and revealing the path that he intends to take. This beautifully illustrates the concept of intention. Initially a thought arises, such as 'I must fetch a cold drink from the fridge', and this thought is then translated into the intention to perform a series of movements that allow you to achieve this objective. This intention in turn is translated into a mental image of the movement or movements.

Some people are not aware of these transient mental images, whereas others make purposeful use of them in training to perform particular tasks. Take the example of a gymnast. When practising a complicated manoeuvre, the gymnast will not only physically practise the movement repeatedly, but will also create a mental image of the movement and repeatedly rehearse the movement in her mind. As she repeatedly *intends* to perform that particular movement, the mental image is continuously reinforced, and becomes much clearer and more readily accessible with every repetition. The intended movement during tai chi allows the practitioner, over time, to both feel and direct the energy or ch'i associated with this intention.

At present, there is no scientific method of measuring ch'i. In a groundbreaking experiment in the 1980s, the French researcher Pierre de Vernejoul injected a radioactive marker into subjects at a number of traditional acupuncture sites.[4] The marker was seen to travel pathways mirroring the classic ch'i meridian lines shown in the diagram (see page 39). The nature of ch'i still baffles the scientific community, and there is still doubt among some scientists about its existence. However, understanding the nature of ch'i is not crucial to understanding meditation, nor its physical and psychological health benefits. Like the psychic abilities that can accompany meditative practice, the ability to perceive and manipulate the flow of ch'i is only a small part of the overall changes evoked by some types of meditation.

A holistic approach to healthcare

There is ample evidence supporting the need for a shift towards a more holistic approach to medicine in the West. In fact, even Hippocrates emphasized the fundamental interrelation between body, mind and environment, and defined health as a 'state of balance'. Although the Eastern approach to healthcare appears to offer advantages over the Western approach in this regard, TCM does not represent a viable substitute for Western medicine. Both traditions bring different tools to our struggle against disease and poor health, and both are valuable in our healthcare system. While many people still resist the integration of Western and Eastern methods, maintaining that one approach is superior to the other, there is an increasing proportion of healthcare providers, in both the West and the East, who see an integrated approach as potentially safer, faster and more effective.

The key contribution of the Eastern approach is the view that meditative practices play a crucial role in healthcare, and this view has been enshrined in ancient healing systems for centuries. But we live in the 21st century, a time of unprecedented scientific knowledge and research. Is there any scientific evidence that meditative techniques actually work and have a measurable effect on the mind and body? In providing this scientific evidence, this book will demonstrate a clear and important role for meditation in both effective patient care and the maintenance of individual health and wellbeing.

Peering beneath the skull

How the brain works

CHAPTER 2

THE HUMAN BRAIN IS AN AMAZING PIECE OF BIOLOGICAL HARDWARE WITH AN ASTOUNDING RANGE OF ABILITIES. It allows us to solve complex tasks, remember events from our childhood, plan our future behaviour, and create breathtaking art, music, architecture and literature. These are feats that cannot be rivalled even by the most sophisticated artificial intelligence created, and the complexity and beauty of the brain's design has yet to be surpassed by any natural or manmade creation.

For many, this is taken for granted; the brain is simply 'grey matter', 'a jelly', that controls and coordinates our thinking and behaviour. And indeed, it is so efficient that many of us run on auto-pilot, responding to external stimuli with pre-programmed reactions. However, on buying a new DVD player, most people would spend a little time looking at the manual to find out how to use the player properly. Similarly, in order to learn how to use our brain effectively, we must take the time to examine how it defines, executes and coordinates our behaviour, and under which conditions it functions the most efficiently. Only by doing so can we start to appreciate and use the potential inherent in this incredible 'jelly'.

In this exploration of the structure and function of the brain, we'll look at the basis of learning and memory, discover the way in which the strength of the connections between adjacent brain cells can be turned up or down like a volume control and find out more about the role that learning and memory play in our perception of the world and our relationship to it and each other. This insight into the brain's processes provides clues about what gives rise to certain modes of behaviour and how, through meditation, we can take advantage of the malleable nature of our brain to release ourselves from the restrictions of our conditioning.

Early views on the brain

The earliest mention of the human brain is in the Surgical Papyrus, discovered by the American Egyptologist Edwin Smith in the late 19th century and thought to be a copy of an even earlier document, written in 3000 BC. The papyrus describes the external surface of the brain in detail and includes records of head injuries. Although they associated damage to the brain with changes in the function of other parts of the body, on the whole the Egyptians did not view the brain as an important organ.

It was Hippocrates, in the fifth century BC, who first recognized the importance of the human brain in defining behaviour. He envisaged it as both a repository of all sensory information, and an interpreter, gleaning meaning from this incoming information. His vision of madness arising from abnormal brain temperatures is undoubtedly simplistic, but his belief in the pivotal role of the brain in the expression of human consciousness remains valid to this day. This vision of brain function was not universally accepted, however – Aristotle, for example, maintained that the heart was the site of human intellect, and that the brain was merely a cooling unit designed to lower the temperature of the blood, 'an organ of minor importance'.[1]

The head/heart debate outlived both philosophers, and it was only in the second century BC that the physician Galen found evidence to support Hippocrates' view. As physician to gladiators, he was able to note behavioural changes in men with head injuries and examine the workings of the human body during the countless surgeries and dissections he performed. His investigations, which led him to suggest that the brain was the seat of human intelligence, inspired the work of the 17th-century English physician, Thomas Willis, considered by some to be the founding father of modern brain science. Willis's *Cerebri Anatome* (*The Anatomy of the Brain*) (1664) provided the Western world with the most detailed insight yet into the anatomy, physiology and pathology of the human brain, and his numerous case studies still remain valid in the light of modern neuroscience.

Picturing the human brain

Countless researchers have focused their efforts on gaining a better understanding of the human brain, but the precise manner in which our brain produces the wealth and scope of human behaviour is still far from clear. As with other avenues of scientific discovery, our methodical scrutiny of the human brain and behaviour has thrown up as many questions as answers. We may now understand the processes through which we see, hear, smell, and even remember an event, for example, but so many aspects of human behaviour still remain a mystery. To what extent is a brain's wiring the result of genetic inheritance and to what extent is it the result of experience? What is the source of human creativity? What inspires us to capture an experience in a painting, a poem or a musical composition? Is consciousness a by-product of brain function? Why do we laugh and cry?

For many years our attempts to unravel the brain's mysteries have been hampered by the tools at our disposal. But with sophisticated imaging techniques, we are now slowly piecing together information, taken from a wide range of different scientific disciplines, to build a more comprehensive picture of the human brain and behaviour.

A neurone and its 'tree' of dendrites

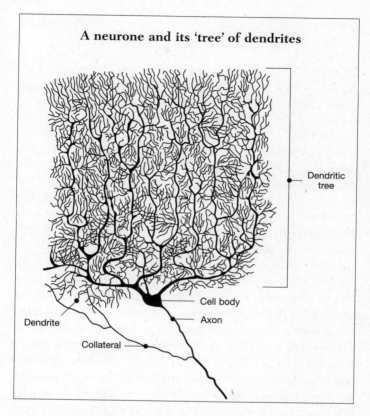

Dendritic tree

Cell body

Axon

Dendrite

Collateral

A network of astounding proportions

The human brain is a reddish grey mass, with the consistency of firm jelly. It weighs on average the same as three bags of sugar, and houses more than 100 billion individual brain cells, called neurones, together with a protective, nourishing and supportive framework of other cells and tissues.

Each neurone has a cell body, which houses its processor, the nucleus. Branching from this cell body are numerous finger-like dendrites. As their name implies, these branch and rebranch, fanning out to extensive, tree-like structures that intertwine with the dendritic trees of other neurones, allowing for many possible points of contact between adjacent cells.

Each neurone makes up to 1,000 different connections with its neighbours and with cells in other regions of the brain. This extensive connectivity allows electrical signals, and thus information, to travel from one brain processing centre to another in a matter of milliseconds. If every single person in the world had access to the internet, the resulting network would still be only a fifteenth the size of the average human brain. That's a lot of connectivity packed into the space of a human skull. Of course, it is impossible for us to count the precise number of connections in the average human brain, but it has been suggested that, at a rough estimate, there are more connections in the human brain than there are stars in the universe. A truly mind-boggling fact.

Each neurone is a miniature processing unit. It receives information from other cells via its connections, processes this information and relays the resulting data to other cells in the network. As the electrical output of a neurone travels through the dendritic tree, it triggers the release of small packets of chemical signals, called neurotransmitters, into the neurone's immediate surroundings. Like a drop of ink released into a bowl of water, these chemical signals diffuse through the fluid between the neurones and bathe all the neighbouring cells. Some neurones release a chemical that has an excitatory effect on neighbouring cells, generating an electrical signal through these cells. Other neurones release an inhibitory chemical that dampens activity in the neighbouring cells, significantly reducing the chances that an electrical signal is generated. In simple terms, therefore, a neurone can either switch on or switch off the activity of other neurones. Activity in the network can be both promoted or dampened depending on the precise cocktail of neurotransmitters released at any one time.

Not only is the level of activity in the network constantly fluctuating, but connections are constantly being pruned and formed. The precise wiring of the brain's network is continually changing, and the network can adopt any one of an unlimited number of different configurations – it is estimated that there are more possible configurations than there are elementary particles in the universe. Not only does every person have an unique configuration, but this configuration is constantly evolving, adapting as a result of experience, ageing and environmental factors such as diet, drugs, stress and disease. If you consider that some of our best computers have processing capabilities on a par with that of a rat brain, it becomes clear that the capacity and complexity of the human brain far exceeds that of any technological devices likely to be created in the foreseeable future. As Isaac Asimov wrote: 'The human brain, then, is the most complicated organization of matter that we know.'[2]

Order in chaos: appearances can be deceptive

When Santiago Ramón y Cajal, who won the Nobel prize for physiology (medicine) in 1904, presented the first microscope-assisted drawings of brain tissue to the scientific community in the early 20th century, the brain appeared as a tangled mess of neurones and supportive elements. On closer examination, however, the morass can be seen to be highly ordered. Within a section of brain tissue, the different cell types are arranged in clear layers, and the cells in each layer share a common orientation.

This precise arrangement of cells in the brain tissue is astounding considering the apparently haphazard manner in which the brain's structure, including its wiring, evolves in the developing embryo. Somehow, most of the time, every neurone ends up exactly where it is meant to be, neatly arranged together with other neurones with the same function. Each neurone appears to be pre-programmed with information about its intended location in the brain and its function, and, under the right developmental conditions, it follows invisible paths

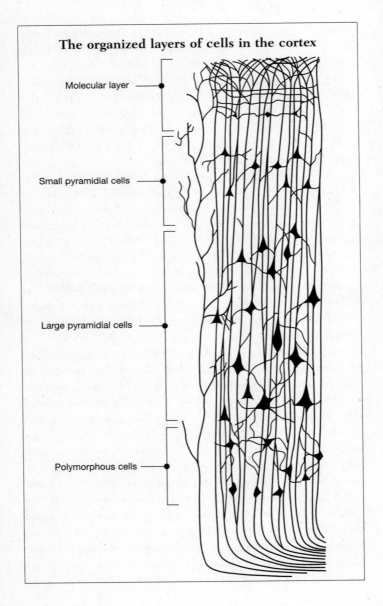

The organized layers of cells in the cortex

Molecular layer

Small pyramidial cells

Large pyramidial cells

Polymorphous cells

through the developing brain until it reaches the appropriate destination. Although the brain has the consistency of a firm jelly, its structure is nothing like this popular dessert. Rather than being a consistent blend of all the ingredients, the brain's different ingredients separate into distinct layers and regions (see opposite), while still maintaining tight links through the countless connections between these layers and regions.

The evolving brain

The complexity of the brain's design is not only apparent in its microscopic structure, but it is also evident at a visible anatomical level. The human brain is built up of three main blocks: the forebrain, the midbrain, and the hindbrain (see page 52).

The oldest part of the human brain, the **hindbrain**, evolved more than 500 million years ago. It closely resembles the brain of a modern reptile and houses neurones responsible for the automatic physiological reflexes that control breathing, heart rate and digestion, and coordinate movement and sense perception.

The **midbrain** is underdeveloped in lower invertebrates, but becomes much more pronounced in primates and humans. It contains neurones responsible for temperature control and the fine-tuning of movement, and it is also the relay centre of sensory information travelling from the body's sensory organs to the forebrain. The midbrain also plays an important role in the limbic system, a group of brain structures thought to be influential in the expression of emotion.

The most evolved part of the human brain is the **forebrain**, which is composed of the cerebral hemispheres, what we mostly commonly think of as 'the brain', and the hypothalamus and thalamus. Although present in all mammals, the cerebral hemispheres are particularly developed in humans. In the last 100,000 years, the weight of the human brain has tripled, and most of this growth has been in the cerebral hemispheres. As

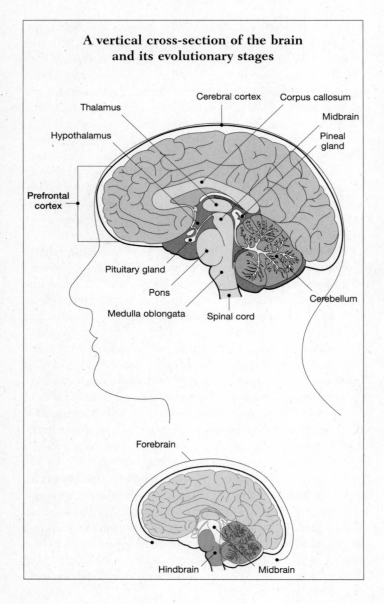

A vertical cross-section of the brain and its evolutionary stages

Thalamus

Cerebral cortex

Corpus callosum

Midbrain

Hypothalamus

Pineal gland

Prefrontal cortex

Pituitary gland

Pons

Medulla oblongata

Spinal cord

Cerebellum

Forebrain

Hindbrain

Midbrain

more and more neural material was packed into the same restricted volume, the brain tissue was forced to fold in on itself, forming deep crevasses, known as fissures or sulci. The neurones of the forebrain control cognitive, sensory and motor function, as well as regulating reproductive functions, eating, sleeping and the display of emotions.

As the human brain evolved, newer brain regions were laid on top of older brain regions, and old but still useful circuitry incorporated into new, more advanced circuitry. The human brain is therefore organized in a hierarchal manner: the oldest parts controlling the more primitive, instinctual behavioural reflexes; the newest parts controlling the more sophisticated cognitive, sensory and motor functions.

A map of human intellect

Historically, determining the specific function of each brain region has been thwarted by the limitations of scientific equipment. Researchers have been reliant on information gleaned from patients with brain damage. However, with the advent of modern imaging techniques, we can now visualize the living human brain and monitor its reactions in real time. We now know which regions 'light up' with activity when a person recognizes a familiar face or tackles a mathematical problem. Slowly, bit by bit, we are piecing together a map of the function of different brain regions: a map of human intellect.

These imaging studies have also shown us that most cognitive tasks, such as problem-solving or strategic planning, activate neurones in more than one brain region simultaneously, or at least in close succession. This property, known as parallel processing, adds a further level of complexity to the organization of the human brain.

Take the example of the processing of a sensory input, such as the sight of a sunset. The sensory input travels from our eyes, through the relay centres in our primitive brain, into the thalamus. From here, it projects out to sensory areas in the cerebral cortex, as the hemispheres are more formally known. Cells in these sensory areas have been specifically designed to respond to specific features of the visual image: its orientation, its colours, its boundaries. At this primary level of processing, we are not consciously aware of any visual image. Information about the lines, shapes and colours of the image have been relayed to the brain, but they have yet to be assembled into the complete image. This occurs at the secondary level of processing.

In the secondary sensory regions of the cerebral cortex, information about colour is integrated with information about the lines and shapes composing the image, and we become consciously aware of the image. (People with damage to these secondary processing areas will believe, for example, that they cannot see, despite retaining the ability to navigate successfully through a room filled with obstacles.)

The image then undergoes a final level of processing, in the tertiary sensory centres of the cerebral cortex. In these so-called association areas (which, as we shall see in the next chapter, play a vital role in the effects of meditation), the image is named and interpreted, assigned emotional value and imprinted in the memory. At this stage not only are we aware of the visual image, but we can also recognize what we are seeing as a sunset and infer the context, such as where we are, or imbue it with symbolic meaning.

This process shows another of the brain's hierarchical structures. The primary level conveys very basic information about the visual image, whereas the tertiary level includes more subtle elements inferred by the image or its context, such as its identity, intention, emotional significance and inherent symbolism. The extensive connections between different brain regions mean that a single sensory input can jolt neurones into life across the brain. Information flows around the brain at a tremendous rate and the whole network realigns itself with the incoming sensory input as it travels upwards through the primary, secondary and tertiary areas.

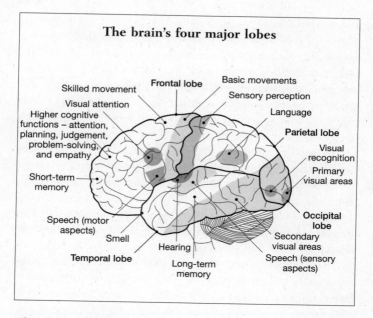

The brain's four major lobes

Skilled movement
Frontal lobe
Basic movements
Visual attention
Sensory perception
Higher cognitive
functions – attention,
planning, judgement,
problem-solving,
and empathy
Language
Parietal lobe
Visual
recognition
Short-term
memory
Primary
visual areas
**Occipital
lobe**
Speech (motor
aspects)
Smell
Hearing
Secondary
visual areas
Temporal lobe
Long-term
memory
Speech (sensory
aspects)

The four lobes

The cerebral hemispheres process all the cognitive skills that we commonly associate with being human: creativity, inspiration, strategic planning, problem-solving, complex emotions, speech and comprehension of language, empathy, insight, memory, visual and audio recognition, and imagination, to name but a few. Although many of these abilities require activity in more than one brain region, some skills, such as visual recognition, for example, can be pinned down to a particular area. It has therefore become possible in recent years to assemble a simplified map of the brain's function. (In the descriptions below, the lobes are discussed in the singular, but there are two of each, one in each hemisphere; although not an exact mirror image of the other, the overall function of these lobes is very similar.)

The frontal lobe

The frontal lobe is associated with cognitive skills such as reasoning, planning, attention, comprehension of speech, skilled and basic movement, the expression of emotions, empathy, short-term memory, motor aspects of speech and problem-solving. Patients with damage to their frontal cortex exhibit a wide range of neuropsychological problems, including impaired concentration, hesitant and interrupted speech, apathy and lack of attention. They can also display a complete lack of inhibition, bouts of euphoria or depression, and a total disregard for the consequences of their behaviour, all of which can lead to socially inappropriate behaviour.

Complex thoughts arise from the activity of neurones in an area of the frontal cortex referred to as the **attention association area**. Located right at the front of the brain, in the prefrontal cortex, the attention association area has extensive connections with the visual parts of the brain and is therefore also involved in generating complex images – it is responsible, for example, for the thoughts and daydreams that we associate with the 'ego'.

In a relaxed mind, thoughts and images spontaneously appear and disappear; different groups of neurones briefly flicker into life in response to incoming sensory information, only to become once again silent when this input has been processed and relayed to other relevant brain regions. In an active mind, on the other hand, certain thoughts and images are given more attention than others; data relevant to the issue at hand are given precedence over irrelevant data. This acts both to reinforce activity in neurones broadcasting relevant information, and to reduce activity in neurones conveying information that's not needed.

The frontal cortex, then, contains the circuitry responsible for screening incoming information for relevant thoughts and images. This becomes especially relevant in meditation.

The parietal lobe

The parietal lobe is associated with sensory perception, language and selective attention. Patients with damage to the sensory areas of a parietal lobe have a distorted perceptual awareness. If someone pinches you surreptitiously, it is normally easy to locate which part of your body was pinched. Damage to the parietal cortex, however, leaves many patients with numbness on one side of the body and an inability to locate the source of incoming sensory information. In one case study, a patient with damage to the back part of both parietal lobes was found to be incapable of filtering out irrelevant visual information. Imagine navigating through a busy train station if you were unable to focus on certain visual objects in order to map your path, unable to block out all the visual distractions in such a landscape.

The ability to navigate around obstacles is also dependent on the formation in the brain of a three-dimensional image of our own body. The area of the parietal lobe responsible for generating this image is often referred to as the **orientation association area**. Not only do cells in this region orientate us in space and time, but their activity also gives rise to our sense of a boundary between 'self' and 'non-self'. This reinforces the sense of an isolated ego; a separate entity that can claim ownership over the thoughts generated by the attention association area in the frontal lobe.

The parietal lobe also contains the **verbal–conceptual association area**. This region of the brain is associated with our finer cognitive skills, such as the generation of abstract concepts from a collection of thoughts, the comparison of concepts, the naming or categorization of objects, and the expression of consciousness through spoken and written language.

The temporal lobe

The temporal lobe is associated with the detection and

recognition of sounds, long-term memory and the sensory aspects of speech. Patients with damage to a temporal lobe cannot understand language or remember words and their definitions.

The temporal lobe also includes structures belonging to the limbic system: the hippocampus and the amygdala. These structures are thought to be involved in generating emotional behaviour, as well as in long-term memory. The temporal lobe also houses the **visual association area**. We have seen that sensory information undergoes three hierarchical levels of processing (see pages 53–54). The final level of processing of visual information occurs in this visual association area. Cells here become active when an object of interest or significance enters the visual field. The extensive connections between these cells and the underlying limbic structures mean that visual images can be endowed with emotional significance, and also that emotions can be directed towards a particular object we see. Damage to the temporal lobe, as well as resulting in the loss of the ability to remember faces, lists of words or groups of visual objects, can therefore also cause emotional flatness.

The occipital lobe

The occipital lobe is associated with the processing of visual information, including visual recognition. Patients with extensive damage to the occipital lobe cannot see, despite the fact their eyes are functioning normally – a condition referred to as cortical blindness. Some patients retain some vision despite extensive damage and are able to correctly guess the location, orientation and movement of a visual object without actually becoming consciously aware they have seen the object. This phenomenon is referred to as blindsight. If only a small region of cortex is damaged, a correspondingly small area of the visual field will be impaired. The larger the amount of spared visual cortex, the better the vision.

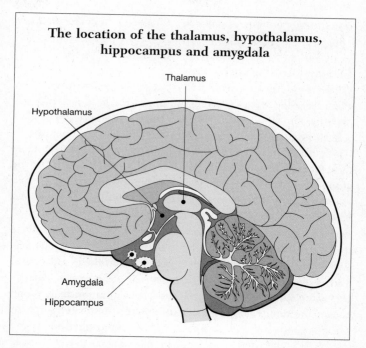

The location of the thalamus, hypothalamus, hippocampus and amygdala

Thalamus

Hypothalamus

Amygdala

Hippocampus

The thalamus and the senses

The thalamus is the gateway for sensory inputs travelling into the cerebral cortex; it relays sensory information to thousands of cortical neurones synchronously. Activity in neural circuits connecting the prefrontal cortex and thalamus are thought to direct our attention, and thus our sensory awareness, towards important sensory stimuli.

The thalamus also relays information between motor coordination centres in the cerebellum, at the back of the brain, and the areas of the brain involved in generating movements and fine-tuning them. Together with the hypothalamus, the thalamus also regulates our level of arousal and awareness. The role of the thalamus in focusing attention, and thus in meditation, will be discussed in more detail in Chapter 3.

The hypothalamus

While areas of the cerebral cortex make possible complex cognitive skills, the day-to-day running of the brain is left to the hypothalamus.

The brain is very sensitive to even subtle changes in temperature, in oxygen supply or in the chemical content of the fluid bathing the neurones. All of these must remain relatively constant to ensure that our brain continues to function optimally. The hypothalamus uses its extensive connections with both the endocrine (hormone) system and the autonomic nervous system to fine-tune the body's internal environment and smooth over any fluctuations triggered by external changes.

The hypothalamus releases hormones into the bloodstream, and also releases hormones that stimulate other glands, such as the adrenal and pituitary glands, to release hormones. Through these hormones, the endocrine system regulates all our physiological processes, from the growth of muscles, bones and tissues to the digestion and metabolism of food, sexual development and function, and even our sleep patterns and appetite.

The autonomic nervous system

The autonomic nervous system comprises two systems that balance each other out. If you encounter a threatening person in the street, the sympathetic nervous system ('fight or flight' response) kicks into action, driving your heart rate and breathing rate up, slowing down any unessential bodily processes, such as digestion, and dilating your pupils. Conversely, if you have just finished a meal and decide to put your feet up, your parasympathetic nervous system ('rest and digest' response) takes the upper hand, slowing your heart rate and breathing, stimulating digestion and increasing blood flow to the skin and organs. These two systems have antagonistic actions – activity in one inhibits activity in the other. As we shall see in the next chapter, this has an important role in meditation.

The limbic system, seat of the emotions

The hypothalamus also forms part of the limbic system. This generates our emotional responses, as well as determining our emotional outlook, motivation, appetite and libido. Although there is still some controversy over the exact brain structures and regions that comprise the limbic system, it is generally accepted that in addition to the hypothalamus, the limbic system also comprises the hippocampus and amygdala. We have two of each, duplicated in the left and right hemispheres (see page 59).

The **hippocampus** is crucial to the formation of memory; it is where our most emotionally charged memories are stored. It receives inputs from all our senses, and assigns emotional value to these inputs. A memory of a traumatic experience in childhood can therefore evoke the same feelings of fear, anger or sadness many years after the event. Although Alzheimer's disease is associated with the loss of cells throughout the brain, nowhere is the loss as great as in the hippocampus. Accordingly, this extensive damage is reflected in persistent and progressive memory loss.

The **amygdala** is an almond-shaped structure deep within the temporal lobe. It has extensive connections with the hippocampus, thalamus and prefrontal cortex, and these connections give it a role in the expression of emotions, such as love, fear and rage. When stimulated, the amygdala produces fearful, aggressive, violent behaviour, so if it is destroyed, animals become tame and indifferent to danger or affection, and humans become incapable of any type of emotional expression.

Forging connections

We have already touched on one of the defining properties of the human brain: that connections between neurones can be switched on or off in line with input received. This flexible, malleable quality is referred to as neuronal plasticity. Every new experience, every incoming stimulus, elicits a change in our brain's configuration. Just as the strength of our muscles depends on how much we use them, the strength of a connection in the brain depends on how often it is used. Our biological networks store a map of data received, and this map is reinforced every time the same data are presented. If, for whatever reason, the pattern of data changes, a new map is stored and reinforced as the data are repeated or re-encountered.

As early as 1890, the renowned psychologist William James highlighted the importance of neuronal plasticity in human behaviour. James described humans as 'mere bundles of habits'. He suggested that all humans acquire behavioural habits through learning in the early stages of their life, and that these habits automate human behaviour in adulthood, so that the majority of the tasks of daily living (from making a cup of coffee to driving to work or even dealing with work issues) are done 'without thinking'.

Our brain is composed of a large number of groups of interconnected neurones. Each small network of neurones is involved in a particular task. Within each network, the neurones frequently communicate with each other and so the connections between them become strengthened. Often it takes very few repetitions of a task before we start to do the task 'without thinking'. The connections between the different neurones involved in performing the task become stronger so that, not only do we complete the task more quickly, but we also do so with minimal thought or effort. The wiring of our brain takes on an imprint of the task. Should a new method be used to perform the same task, new neurones will be recruited and old neurones decommissioned; the old imprint erased and a new imprint formed.

This ongoing process not only provides an insight into the way in which we retain new experiences and overwrite old experiences so efficiently, but it also provides the key to successful psychotherapeutic change: frequent and repeated reinforcement of the new pattern of behaviour, together with deliberate avoidance of the old pattern.

The plasticity of our brain connections underlies learning and memory, and is a key to unravelling some of the mysteries of human behaviour. Our brain has been designed to encode our behaviour as if it were 'set in stone' – once we learn as a child to ride a bike, we never forget. However, it has also been designed to erase this information and replace it with new behaviour when necessary. New behaviour arises from a conscious process of learning: the repeated presentation of new experiences or stimuli moulds new neural circuits; and the repeated avoidance of old experiences or stimuli erases old neural circuits.

Teaching a dog new tricks

In the early 20th century, the Russian physiologist Ivan Pavlov performed some groundbreaking experiments. The presentation of food triggers a salivary reflex in dogs. Pavlov discovered that, if he placed a morsel of food under the tongue of a dog a certain number of times, the dog would start to salivate on the sound of him entering the laboratory rather than on the presentation of the food itself. The dog had learned to associate the sound of Pavlov's steps in the corridor with the imminent arrival of a tasty morsel. Sounding a bell immediately before providing food had a similar effect; the dog learned to associate the sound of the bell with food, and salivated on the sound of the bell. This demonstrates one of the simplest methods of learning: classical conditioning.

Classical conditioning can also be seen at work in humans. Some of our behavioural reflexes are unconditioned – that is, they are inbuilt imprints we have inherited. One example is the dilation of the pupils of our eyes in low light. Other reflexes are learned at an early stage of our lives: we touch a hot iron, we feel pain and rapidly withdraw our hand from the iron. Since the act of touching the hot iron can be seen to precede the pain, this action is determined to cause the pain. Acknowledging this causal relationship conditions us to associate touching a hot iron (cause) with pain (effect), so modifying our future behaviour.

This simple form of cause–effect conditioning forms the basis of much of our behaviour even in adulthood. Each experience leaves an imprint in our brain's wiring, and collectively these imprints form a conceptual map that guides our behaviour. This map is constantly changing as we gather new experiences and therefore learn new causal relationships. Mostly, the greatest amount of learning, and change, happens when we are young. As children, we are incredibly inquisitive about our surroundings and ourselves. We learn by trial and error, roaming around our new world and investigating it with the enthusiasm of the most avid scientists. In these early years, this learning is crucial to our survival. All our experiences are incorporated into our rapidly evolving conceptual map, and this framework begins to guide our behaviour. Our family, our peers, our society and our culture all strive to inculcate in us appropriate reactions to specific circumstances. This conditioning also moulds our conceptual map. By adulthood, our conceptual map is well established and sometimes even rigid. Many of the experiences contributing to this framework have been endlessly reinforced, and as a result the encounter of similar experiences elicits an immediate unconscious, knee-jerk reaction.

What makes a strong connection?

When two neurones communicate with each other repeatedly, in the presence of other reinforcing signals from other neighbouring neurones, they appear to 'learn' how to communicate with each other more efficiently. This observation, credited to the American psychologist Donald Hebb, forms the foundation of a now longstanding theory of human learning and memory. Hebb's rule states: 'When an axon of cell A is near enough to excite a cell B and repeatedly or persistently takes part in firing it, some growth process or metabolic change takes place in one or both cells such that A's efficiency, as one of the cells firing B, is increased.'[3] In other words, the effectiveness of communication between two neighbouring neurones can be both up- and down-regulated; fine-tuned rather like turning the volume control on a stereo.

In the example of Pavlov's dogs, communication between the neurones controlling salivation and those involved in the perception of food is innate, instinctual. The dog doesn't have to learn to salivate when it sees food; it is an inbuilt physiological response. However, communication between the neurones involved in the perception of the sound of a bell and those involved in the perception of food is not innate. The dog only learns this association if the unconditioned stimulus (salivation) comes to be associated with the conditioned stimulus (the bell). The conditioned stimulus has a much greater impact on the neurones controlling salivation since these neurones are already active in response to the unconditioned stimulus (salivating at the appearance of food). The efficiency of neurone communication, and therefore the strength of the connection between neurones involved in salivation and those responding to the sound of the bell, has therefore been increased.

This type of learning underlies associative memory: a smell associated with a particular person; a song associated with a particular event; a feeling associated with a particular location. A camera is only capable of capturing a single aspect of an experience, the visual appearance, but the human brain can capture sight, sound, smell, tactile feeling, taste *and* emotion, or the sense that the memory has emotional significance.

Those aspects of a particular experience that evoke the strongest emotional response are remembered most vividly. In the same manner as a triage nurse in Emergency examines a series of patients and then refers them for treatment in order of highest medical priority, the brain examines all aspects of a particular experience and prioritizes those with the greatest

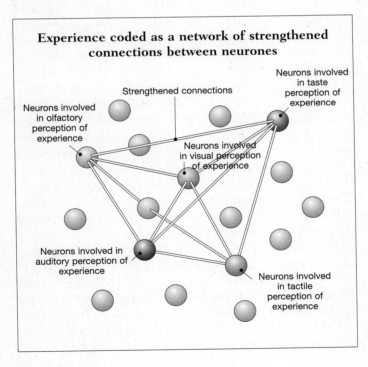

Experience coded as a network of strengthened connections between neurones

Neurons involved in taste perception of experience

Strengthened connections

Neurons involved in olfactory perception of experience

Neurons involved in visual perception of experience

Neurons involved in auditory perception of experience

Neurons involved in tactile perception of experience

emotional relevance to the individual. Therefore an experience is encoded in the brain in the form of a network of strengthened connections. An example of such a network is illustrated opposite.

More than the sum of its parts

Studying the detailed structure of the brain and its neural connectivity is bringing us greater understanding of many of its processes, but much of what the human brain is capable of eludes this type of analysis. Professor Paul Davies, theoretical physicist and cosmologist now working at the Arizona State University, provided a useful analogy: we are composed of atoms in much the same way as a building is composed of bricks, a book is composed of words or a melody is composed of notes. All of these, however, are clearly more than just a collection of their essential building blocks. It is only at the collective level that certain properties emerge, such as an architectural style in the case of a building, a plot in the case of a book, a theme in the case of a melody. Similarly, many human qualities – inspiration, creativity and empathy, to name just three – arise from complex interactions and relationships and cannot be investigated or appreciated by dissecting the brain into its constituent parts. Instead, a more holistic approach is needed.

Quantum revelations

This shift in approach was initiated by the discoveries, in the early part of the 20th century, of a pioneering group of theoretical physicists. It was already well established that all matter was composed of atoms and, with the British physicist Joseph Thomson's discovery of the electron in 1897, many physicists believed that the atom was composed of even smaller particles. Researchers such as Einstein, Planck, Bohr, Heisenberg, Schrödinger, Pauli and Bohm set about to describe the behaviour of these sub-atomic particles, but they came to the realization that the strict, mechanical laws that were so evident at the scale of our everyday physical world could not be applied at the sub-atomic level: the behaviour of particles at this quantum level proved unpredictable and their experimental observations were at best confusing, and at worst nonsensical. There is no one, universally accepted interpretation of these experiments, but it is widely accepted that quantum theory revealed both the principle of complementarity and the interrelatedness of all matter.

The principle of complementarity

At a quantum level, matter displays both particle-like and wave-like properties. Take the example of light. Depending on the experimental apparatus used, light can be detected both as a wave (light wave) or as a particle (laser beam). Niels Bohr proposed that these two phenomena were complementary descriptions of the same reality. Taken individually, each description is only partially correct, but taken together a more complete vision of reality emerges.

The interrelatedness of all matter

Quantum particles cannot be described or defined in terms of their individual properties, but only through their interactions with other particles: 'Quantum concepts imply that the world acts more like a single indivisible unit, in which even the "intrinsic" nature of each part (wave or particle) depends to some degree on its relationship to its surroundings.'[4] An implication of this for scientific investigation is that even subtle changes to the experimental set-up can evoke changes in the properties exhibited by sub-atomic particles – a sub-atomic particle cannot be isolated, or considered independently from, the processes involved in its creation, measurement and destruction. The scientific experimenter, who was until now considered to be an independent and objective observer of reality, can influence the outcome of an experiment by the choice of experimental apparatus or measurements taken.

How do these findings relate to our vision of how the brain works?

The quantum brain

In his book *The Quantum Brain*, Jeffrey Satinover proposes that all life, including the human brain, has evolved to amplify quantum effects, and therefore take advantage of them. Effects at a quantum level are translated into subtle changes in the shape of the proteins that comprise our body. This signal is then amplified by many orders of magnitude: changes in a single protein can trigger a change in the array of proteins within which it is located; these changes in turn instigate changes in the cytoskeleton of the cell, resulting in changes in the structure and therefore function of the whole cell; changes at the cellular level are then transmitted to all the cells within a particular network; and finally, these network changes can bring about changes in the overall functioning of the brain.

This means that brain function should be examined both in terms of its particle-like properties (reductionist approach; the level of individual neurones) and its wave-like properties (holistic or systems approach; the level of networks of neurones). Taken individually, the two approaches provide an incomplete description of brain function, but when combined and viewed as complements of each other, they allow us a more complete vision of how the brain functions.

The left and right hemisphere: complementarity at work

With respect to sensory processing and motor function, the left and right cerebral hemispheres have a similar function: the right hemisphere controls movement and receives sensory input from the left-hand side of the body, whereas the left hemisphere controls motor and sensory function on the right-hand side of the body. However, in other respects, the function of the two hemispheres is asymmetrical. This is a unique feature of the design of the human brain.

Since the left hemisphere contains the brain regions associated with both written and spoken language, it is often thought to be the dominant hemisphere. However, the right hemisphere contains regions that are associated with the comprehension and generation of emotional inflection to language. Cooperation between the hemispheres is therefore important: the left and right hemispheres each produce a partial but complementary description of the world.

Generally speaking, the left hemisphere is associated with analytical, rational and logical processing, whereas the right hemisphere is associated with abstract thought, non-verbal awareness, visual–spatial perception, and the expression and modulation of emotions. For example:

Linear versus holistic processing
Imagine tackling a problem. The left hemisphere acts to identify all the different components of the problem, assemble them into a logical order and then draw conclusions. The right hemisphere generates the 'big picture' and then works backwards from the 'answer' to fill in the details.

Symbolic versus concrete processing
Imagine learning to cook a new recipe. The left hemisphere can interpret symbols, such as words or mathematical formulae, and therefore learns by rote. The right hemisphere learns through hands-on experience or by watching someone cook the dish.

Verbal versus non-verbal processing
Imagine giving directions. The left hemisphere can verbally describe directions in a logical manner: the first turning on the left, right at the fork. The right hemisphere can only process directions if they are given in the form of a visual map.

Further insights into the relative roles of the two hemispheres came from split-brain surgical operations performed in the 1960s, in which the connections between the two hemispheres were severed, leaving the two hemispheres essentially isolated from each other. The right hemisphere appears to 'know' an object in terms of its function or purpose; the left hemisphere 'knows' an object in terms of its name or category. Neither one of these two modes independently provides all of the information about the object.

Suppose you're inspecting a new object, some oddity collected on a friend's travels. The left hemisphere begins to examine the object, noting how it has been made, what it is made of and whether there are any distinguishing marks. From these clues, the left hemisphere might hazard a guess at what the object might be or how it might be appropriately categorized. This gives rise to a partial description. The right hemisphere notes any function inherent in its design or any clues as to what it may have been used for. The right hemisphere puts this information in the context of what it already 'knows' about the world and then offers a suggestion for its function. This provides a different, but also partial description of the object. A deeper understanding of the object arises from the coordinated activity of both the left and the right hemisphere. By working as a team, sharing this information, the two hemispheres can generate a more complete description.

Interestingly, more recent experiments in split-brain patients have also revealed that the right hemisphere provides a more 'truthful' description of a particular experience. Because the left hemisphere is involved in logical and analytical thought, it attempts to extract meaning, to impose order. This pattern-seeking behaviour can lead us to extract meaning from something that is purely random. The left hemisphere is very inventive and creative; it is the storyteller, our inner narrator. The intuitive and abstract approach of the right hemisphere, on the other hand, is capable of accurately capturing the whole 'present-now experience', with the minimal amount of filtering or interpreting. When memories are accessed through processes occurring in the right hemisphere, the content of these memories are therefore often fuller and more realistic – you relive the memory as it happened rather than as an interpreted, rationalized version.

In the next chapter we shall examine in further detail the significance of these differences in the context of the neuropsychological basis of quests such as that for 'the meaning of life'.

The brain and interrelatedness

The brain is more than the sum of its parts. To take the analogy from page 67, the words in a book can usefully be studied individually, but it is only by looking at their interrelatedness that a book's 'emergent properties' – the plot, the style, the message – become apparent and can be understood. At the individual word level such emergent properties are destroyed, as is our ability to investigate them.

Similarly, activity in the brain can be viewed at the level of single synapses – observing how the connections between brain cells change over time – but also at a much higher level. Human characteristics such as inspiration and creativity are not determined by activity in a distinct and identifiable region of the brain; they emerge through the coordinated behaviour of large collections of neurones, spread over different regions of the human brain. If we are to gain a better understanding of some of our more elusive cognitive skills, our brain's emergent properties, we must examine them at the level of the whole brain.

With the advent of non-invasive imaging techniques, such as Magnetic Resonance Imaging (MRI), we can study and measure activity across groups of neurones, even across the whole brain, and watch it in real time. This has revealed the different areas of the brain involved in specific tasks, the sequence in which activity in different brain regions changes during a particular task, and the impact of lifestyle factors and disease on brain function. Combining this new knowledge with detailed studies of individual neurones, their connections and their chemical signals, offers a more complete picture of how the brain works and responds, and is allowing us slowly to unravel the neural basis of human experience.

In recent years a number of researchers have proposed that the brain stores information holographically – as a 3D representation. The 'phantom limb' phenomenon is an example of this. Patients who have lost an arm or a leg, for example, report still feeling sensation in the absent limb, or act as if their lost limb were still present – it appears that a sense of the complete physical body is still there despite the lack of any sensory information actually coming from the missing limb. Our brain creates our personal reality and projects it out as a holographic representation. Limbs can be felt when actually absent, because the *holographic representation* remains intact, despite the absence of a limb.

The assertion that this holographic storage of information goes beyond phantom limbs is supported by two major experimental observations. First, when a particular brain region is damaged, another region of the brain can assume the damaged region's responsibilities, especially if the damage occurs at an early stage of the person's life. Secondly, huge portions of the cerebral cortex can be removed or damaged before consciousness is lost. Only when the large majority of the cortex is out of action, as in particularly severe cases of epilepsy, for example, does the emergent property of consciousness disappear. Emergent properties arise from the global, dynamic functioning of the brain, and it is plausible that this global functioning is somehow encoded at the level of each individual brain component.

Subjective values

In the study of the human brain and behaviour, the loss of scientific objectivity is especially relevant. Just as the exact set-up of a piece of apparatus can determine the outcome of an experiment, so the exact configuration of our conceptual map at a particular time determines our perception of a particular experience. No experience can be purely objective; all our experiences are filtered and interpreted by our brain, giving them a subjective gloss however hard we may try to avoid it. This loss of scientific objectivity refutes our sole dependence on rigorous, scientific experimentation and opens the door for other experiential sources previously disregarded for their presumed subjectivity.

Subjective accounts of the brain's emergent properties have always been interpreted with caution, since they cannot, correctly, be dissociated from the individual person. However, this experiential approach does offer glimpses of our brain's emergent properties not afforded by traditional scientific method. We are hard-wired to extract meaning and learn from our experiences, and to encode this learning for our entire lives. Furthermore, we are hard-wired to operate in one of two modes of knowing or thinking (i.e. left-brain and right-brain thinking), and can therefore both ask pertinent questions and devise appropriate answers. What we learn through these glimpses into our own minds can guide our future scientific investigations into phenomena such as consciousness, creativity . . . and the meditating mind.

Meditation and mystical experiences

CHAPTER 3

ADVANCES IN TECHNOLOGY MEAN THAT WE CAN NOW RECORD THE
ACTIVITY OF SINGLE BRAIN CELLS, produce images of activity in
different brain regions, and pin down the role of individual
chemicals in the brain in generating the wealth of human
behaviour. This technology, together with the map of the
human intellect provided by imaging studies, allows us to delve
deeper into the inner working of the human brain.

In recent years, we have seen a proliferation in neuroscience
research focusing on the brain's involvement in the attainment of
altered or higher states of consciousness through meditation. By
linking specific changes in the brain during meditation with the
subjective experiences frequently associated with meditation, we
can begin to see how and why meditation works.

Characteristics of religious and mystical experiences

Religious and mystical experiences are characterized
by optimism and are associated with a number of
important features.

- First, they defy expression in terms that can be
  understood by someone who has not had, at some
  point, a comparable experience.
- Secondly, they can be viewed both in terms of a
  different state of feeling and a different way of
  knowing. Ordinary knowledge is seen to be indirect
  and intellectual, whereas mystical knowledge is
  viewed as direct, intuitive and immediate – those
  experiencing them have absolute conviction of the
  'truth' of these experiences.
- Thirdly, they are usually short-lived and therefore
  must be viewed as transient events. However, not
  only can regular meditation increase the frequency
  with which these experiences occur, but the changes

seen in the brain during these experiences can also spill over into normal waking life and sleep.

- Fourthly, they have an element of passivity, in that they often feel as if they have been presented as a gift, requiring no personal effort or will. In fact, the superimposition of goals or motives on to the practice of meditation can lessen the likelihood of these experiences happening.

- Finally, those undergoing religious or mystical experiences are struck by the sense that the ego is not the true self, but a limited and constrictive fabrication of the mind. This can be associated with feelings of 'unboundedness' and the 'loss of time, space and body sense'.

The brain's involvement in mystical experiences

The first indication of the brain's involvement in religious and mystical experiences was the observation that patients with temporal lobe epilepsy often had hallucinations with a mystical or religious content and reported seizure-induced feelings of religious ecstasy. In some cases, epileptic patients even reported spontaneous religious conversions. There is some evidence that a number of mystics from the past may have had temporal lobe epilepsy. St Teresa of Ávila, for example, is reported to have had visions, chronic headaches, transient loss of consciousness and other symptoms attributable to temporal lobe epilepsy.

Arousal in the temporal lobe and the 'God machine'

In 1997 the neuroscientist Vilayanur Ramachandran presented a paper at the Society for Neuroscience annual conference in New Orleans entitled 'The Neural Basis of Religious Experience'. Ramachandran and his team had studied patients with temporal lobe epilepsy. These patients were found to experience frequent microseizures in their temporal lobe. In many cases, these microseizures were associated with religious experiences or fanaticism.

In order to investigate this phenomenon further, Ramachandran measured the level of arousal in the brain when patients were shown 40 different words, including violent words, sexual words, neutral words (such as 'wheel') and words with religious connotations. Results were compared with two control groups: a religious control and a non-religious control. Ramachandran and his colleagues found that patients with temporal lobe epilepsy showed much greater arousal when presented with religious words than non-religious words. The religious control group, on the other hand, were aroused by both religious and sexual words, while the non-religious control group were only aroused by the sexual words. These observations led Ramachandran to suggest that patients with temporal lobe epilepsy were in some way more sensitive to religious words, and that every time a patient had a seizure, their interest in religion became strengthened.

Ramachandran's paper prompted a media frenzy. The media proclaimed that a 'God module' had been found in the brain and, as we will see in Chapter 4, this was to spark a heated debate that has yet to be resolved. Some people believe these findings provide the definitive scientific proof of the existence of God. Others believe they provide the definitive scientific proof that God is a product of brain function. However, on face value, Ramachandran's research merely suggests that the temporal lobe is, in some way, involved in religious experiences.

Ramachandran's research led an American neuroscientist, Michael Persinger, to suggest that religious experiences were the result of microseizures in the deep structures of the temporal lobe. Persinger proposed that these microseizures gave rise to the entire spectrum of religious behaviour, from 'early morning highs' through to religious conversion or even extreme religiosity. Although these microseizures appear to be produced by a large variety of different stimuli, they are most notably triggered by personal life crises and near-death experiences. They can be experienced by healthy individuals as well as epileptics, and should not therefore be viewed as merely a pathological process in the human brain. Persinger suggests that an individual's susceptibility to these microseizures depends on the lability of the temporal lobe – that is, how prone it is to change. Individuals with a high temporal lobe lability appear to be more likely to experience microseizures, and so are more likely to have religious experiences. Interestingly, people with higher temporal lability were also found to be associated with predominantly right-brained – intuitive rather than rational – thinking.

In order to test this theory, Persinger designed a machine that was capable of producing microseizures in an experimental subject. This machine, little more than a motorcycle helmet with magnets, delivered a weak magnetic field over selective portions of the brain, effectively triggering microseizures in the underlying brain tissue. This technique was reported to trigger religious and mystical visions in healthy subjects. In an article for the *Washington Post*, Persinger stated that 80 per cent of people tested in his laboratory reported sensing an entity either behind or near them when wearing his helmet. Stimulation of the right temporal lobe was reported to produce more pleasant experiences than those evoked by stimulation of the left temporal lobe.

In John Horgan's book, *Rational Mysticism*, Persinger states that his helmet – sensationally named the 'God machine' by the media – has, as yet, not been able to induce a religious or mystical experience in its creator. Persinger explains this lack of effect on his generally sceptical and scientific state of mind. As higher levels of temporal lobe lability are associated with right-brained thinking, individuals who are predominantly left-brained thinkers can be assumed to have lower levels of temporal lobe lability and therefore a reduced likelihood of a religious or mystical experience on temporal lobe stimulation.

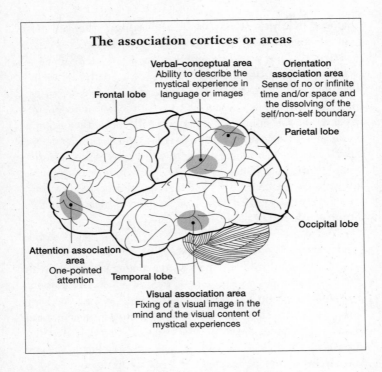

The association cortices or areas

Verbal–conceptual area
Ability to describe the
mystical experience in
language or images

Frontal lobe

Orientation
association area
Sense of no or infinite
time and/or space and
the dissolving of the
self/non-self boundary

Parietal lobe

Occipital lobe

Attention association
area
One-pointed
attention

Temporal lobe

Visual association area
Fixing of a visual image in the
mind and the visual content of
mystical experiences

Patterns of brain activity during mystical experiences

Persinger's results have been met with considerable scepticism. The stimulation of certain areas of the brain to evoke specific experiences is undoubtedly not the best way of investigating these experiences. These artificially evoked experiences can rightly be viewed as artefacts of brain function. Accordingly, neuroscientists Andrew Newberg and Eugene d'Aquili set about observing mystical experiences evoked by one of a number of different spiritual techniques or disciplines, including meditation and centering prayer.

The rare and fleeting nature of these mystical experiences makes their observation under controlled conditions in the laboratory difficult. However, Newberg and d'Aquili were partially able to overcome this obstacle through a deceptively simple technique. After inserting an intravenous tube into the subject's arm, each subject was asked to meditate as they would normally do. All subjects used an image as a point of focus, usually an image of religious significance. After an average of 60 minutes, when the subject felt that the boundary between themselves and the image was dissolving, they tugged on a piece of string, thereby triggering the release of a radioactive tracer into the blood system. The larger the blood flow to a particular region of brain, the greater the amount of tracer bound to the brain tissue. Using a specialized brain-imaging technique, the blood flow in different regions of the brain, and hence the activity in these regions, could then be visualized.

Through their research, Newberg and d'Aquili have shown that mystical experiences are associated with specific patterns of brain activity not limited to the temporal lobes. Whereas spontaneously evoked religious experiences appear to involve circuitry that lies within the temporal lobe, mystical experiences evoked by meditation appear to involve circuitry throughout the entire brain.

We saw in Chapter 2 that each hemisphere contains four association areas: the orientation association area (parietal lobe), the attention association area (frontal lobe), the visual association area (temporal lobe) and the verbal–conceptual association area (parietal lobe). Newberg and d'Aquili propose that a sequence of changes in the activity in these areas gives rise to the unfolding of the different stages of mystical experiences and their associated features. These processes occur first in the right side of the brain, but spread to the left through the extensive connections between the two hemispheres, leading to a similar chain of processes in the left side of the brain.

The unfolding of this sequence of changes in the brain can

lead to mystical experiences, and it is these processes that have been harnessed by numerous different contemplative disciplines to evoke and enhance human development.

The orientation association area

This area in the parietal lobe establishes a three-dimensional sense of 'self', creating a boundary between 'self' and 'non-self' that orientates us in physical space and time. The construction of an arbitrary boundary line between 'self' and 'non-self' is clearly essential to our ability to interact with our physical world. Our ability to move around without bumping into things, for example, requires the formation of a mental map that contains details about the environment in terms of how far away objects are and what relevance they have. In many ways, therefore, the sense of 'self' can be viewed as an artefact of our interaction with the physical world.

Newberg demonstrated that during transcendental experiences there is a dramatic *decrease* in activity in the orientation association area. This has the effect of dampening the sense of time and/or space and of dissolving the self/non-self boundary.

The attention association area

Across the different contemplative disciplines, various techniques are used to focus the mind to single-pointed attention, including mantras, chanting, images and repetitive movement. All these techniques trigger activity in the attention association area in the prefrontal cortex. The neurones in this area of the brain are associated with goal-directed behaviour, both physical and psychological. Navigation through a crowded train station, for example, requires activity in this part of the brain, as we filter out redundant sensory information and focus only on those elements that can guide our path through the station. Similarly, in order to examine a thought properly or come up with a plan of action, we need to focus on the issue at hand and filter out redundant thoughts. By giving the mind a single focus, someone meditating is using innate neural circuitry to filter out external noises and internal thoughts.

The visual association area

The neurones in the visual association area of the temporal lobe are involved in assigning relevance to specific objects or processes in the visual field. Their function is therefore closely tied with that of the neurones in the attention association area, and the use of a visual image or object to focus the mind to one-pointed attention requires activity in both association areas.

As well as being involved in the processes leading to mystical experience, the visual association area is also undoubtedly responsible for the visual content of such experiences. We have already seen that the temporal lobe houses structures associated with memory and emotion. The visual association area therefore has access to a vast repository of memories and emotions to give the mystical experience its visual flavour and emotional significance. The extensive connections between the visual association area and the brain systems involved in memory and emotion also explain why objects of perceived emotional or spiritual significance are particularly effective at focusing the mind.

The verbal–conceptual association area

Our ability to communicate abstract concepts in language relies on activity of neurones in the verbal–conceptual area, which lies at the border of the parietal and temporal lobes. During meditation, there is a *decrease* in activity in this area. This leaves the meditator unable to describe adequately the mystical experience in terms of language, images or any other tool at their disposal – which would explain why mystical experiences cannot be categorized or conceptualized; they can only be experienced.

The process of meditation

There are two main types of meditation: passive and active. Passive meditation involves a widening of the attention to an all-embracing focus, whereas active meditation involves a narrowing of the attention to a one-pointed focus. As these two meditative processes evoke a slightly different set of processes in the brain, we shall look at each in turn.

The following description of these two processes is a simplified version of that given by Newberg and d'Aquili.[1] For each stage of meditation, the changes in brain activity are illustrated with a case study of a hypothetical meditator alongside a neuroscientist's commentary. It should be emphasized, however, that experiences during meditation differ considerably from one individual to the next. Experience during meditation is highly personal and subjective, and the way in which certain indescribable elements of the meditative experience are captured in words will always be inherently limited and somewhat unsatisfying. These descriptions should therefore not be taken to be representative of all meditative experiences, or even a typical meditative experience.

Passive meditation

The meditator begins with the intention to clear his mind of thoughts. This intention is reflected in an increase in activity in the attention association area. As he slowly quietens his mind, in some cases by focusing on the gap between thoughts, there are further increases in activity in the attention association area. At the same time, activity in the frontal cortex regions surrounding the attention association area decreases. This is the result of focused attention and reflects a filtering out of all information that is not deemed important. Attention is drawn to the present-now experience, which triggers a shift to right-brained activity, as attention is predominantly a right-brained function. This shift from 'intellectualized' left-brained thinking is a further explanation of why the experience cannot be

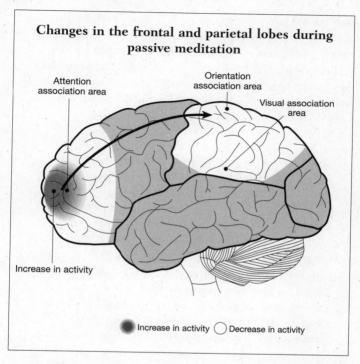

Changes in the frontal and parietal lobes during passive meditation

Attention association area

Orientation association area

Visual association area

Increase in activity

● Increase in activity ○ Decrease in activity

described or analysed: the right brain does not have the ability to categorize and analyse the experience; it intuitively 'feels' it.

At the same time, the meditator also becomes less aware of sensory information stemming from his external environment, and therefore less aware of his orientation in space and time. Imagine sitting in a room, with your eyes closed, body relaxed, allowing your mind to wander. The information that our brain uses to form an image of our body in our mind is not available under these conditions: there is no visual input, no movement, no interaction with the world. Slowly, therefore, our body image begins to fade and we become less aware of our external environment. This dissolving of the self/non-self boundary is reflected in a decrease in activity in the right parietal lobe. Not only does this have an impact on activity in the right orientation association area (leading to a loss of sense of space and/or time), but it also has an impact on activity in the right verbal–conceptual association area, leading to an inability to convey the experience efficiently through language. This sequence is shown in the illustration on page 85.

Meditator: Periodically, throughout the day, I become consciously aware of the fact that my mind is not focused on the now – instead, I'm fretting about some potential experience in the future. As I become aware of this, I bring my mind to focus on whatever I'm doing. This might be writing a letter or something as mundane as doing the washing up. With this intention my body relaxes and there is a conscious decision to focus my attention on experiencing the moment. Thoughts spontaneously pop into my head – something I forgot during a recent trip to the supermarket, a comment from a work colleague earlier in the day – but after acknowledging them I just return to the present-now. Slowly, over time, the gap between these interrupting thoughts gets longer and it takes no effort to focus my attention on the task at hand.

Neuroscientist: At this point, there is an increase in activity in the attention association area and a decrease in activity in the surrounding areas of the frontal cortex. From time to time there are short bursts of activity in neurones in the frontal cortex. This reflects random thoughts arising and then dissipating. As the activity in the attention association area increases even further, with persistent one-focused attention, the short bursts of activity are eventually dampened and thoughts become more infrequent and less interrupting.

Meditator: My awareness of my surroundings recedes into the background. At times I lose myself in the present-now, and time passes during which I'm not aware of my surroundings, not aware of my body or the ache in my back that was troubling me earlier, not really aware of where I begin and where I end. I feel a union with something much greater than myself, something much more expansive than my restricted and rigid sense of self.

Neuroscientist: A decrease in activity in the orientation association area has occurred. Through one-pointed focus, the individual effectively filters out any so-called redundant information, including information from the sensory elements that build up an internal body image. As a result, the body image becomes blurred, and the boundary between body and everything else also becomes blurred. This gives rise to the sensation of unity with something that is greater than 'self'.

This chain of events is thought to result in the activation of two important structures in the limbic system: see the illustration on page 88. There are extensive connections between the parietal lobe's orientation association area and the hippocampus. A decrease in activity in the parietal lobe results in the activation of the hippocampus (Step 2), which in turn stimulates the amygdala. These two structures are responsible for assigning emotional significance to our experiences. The

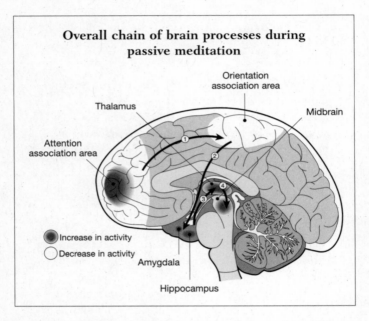

Overall chain of brain processes during passive meditation

Orientation association area

Thalamus

Midbrain

Attention association area

① Increase in activity

○ Decrease in activity

Amygdala

Hippocampus

activation of the hippocampus conveys emotional significance of the experience and imprints the emotionally charged experience in our long-term memory.

Under normal conditions, the activation of the amygdala also registers the emotional significance of the incoming sensory information, and generates an appropriate autonomic, behavioural, motor, hormonal and pain-suppressing responses. Under meditative conditions, however, the activation of the amygdala confers emotional significance on the *lack* of incoming sensory information, and generates appropriate physiological responses to this lack of sensory information. Through actions on the hypothalamus, the amygdala modifies the activity of the autonomic nervous system (Step 3). First, a blissful, peaceful state arises from maximal activation of the parasympathetic (relaxation) nervous system and then, as different neural, hormonal and other triggers swing in, there is maximal activation of the sympathetic (arousal) nervous system, producing a mentally clear and alert state. Physiological effects, such as changes to breathing rate, heart rate or blood pressure are the result of the amygdala's effect on midbrain structures that control these functions (Step 4).

With sustained attention, this activity in the right hemisphere eventually spills over to the left hemisphere, owing to the extensive connections between the two hemispheres. Both the left and the right orientation and verbal–conceptual association areas are therefore switched off. A lack of activity in the *right* orientation association area gives rise to the sense of unity and wholeness, whereas lack of activity in the *left* orientation association area results in the dissolving of the self/non-self boundary. This, and the lack of both the left and right verbal–conceptual area leaves the meditator struggling to find a way of describing the experience to others.

Meditator: As my awareness of the unity that lies beyond my restricted sense of self grows, my whole body appears to respond. A wave of bliss washes over me, like the sun emerging from a cloud and bathing me in light. I feel tremendous peace and union with all. Sometimes, at this point, an image or memory might appear and, with it, strong emotions. This can be enough to pull me back to myself as I delight in the image or replay the memory. But sometimes I manage to merely acknowledge the image or memory, storing it away for future examination, and then return to the present-now experience. Other times, I remain in this peaceful state for an indeterminable amount of time before emerging from my meditation. Sometimes, I become aware of a clarity of mind I don't normally experience in everyday life. From time to time, this clarity provides me with an insight or the answer to a question I'd been thinking about earlier in the day. I just seem to know the answer without being sure of the source. It's hard to explain. It appears to defy logic.

Neuroscientist: The decrease in activity in the orientation association area produces an autonomic nervous system response. This gives rise to the feelings of bliss that accompany the dampening of activity in both the right and left parietal lobe. The right-brain function stemming from meditation gives the individual access to right-brained 'big picture' thinking and right-brain vivid and accurate memories. The lack of activity in the left parietal lobe explains why knowledge gained during meditation is seen to stem from something greater than 'self'.

Active meditation

The meditator begins with the intention to focus her mind on a single object, image or mantra, or on her breath. This point of focus, whether it is something the meditator is looking at or is visualizing in her mind, causes an increase in activity in the attention association area, reflecting an increase in attention. This focused attention is associated with a corresponding decrease in the number of redundant thoughts. There is also an increase in activity in both the visual and orientation association areas, both of which are essential to our ability to visually fixate on an object. This increased activity allows the meditator to fix the image in her mind and, eventually, hold it there with minimal effort (see the illustration opposite).

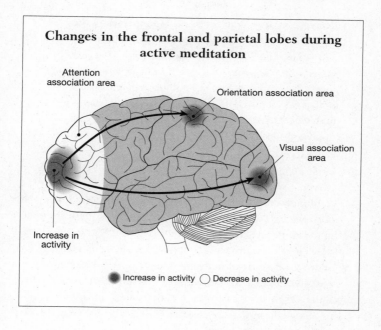

Changes in the frontal and parietal lobes during active meditation

Attention association area

Orientation association area

Visual association area

Increase in activity

● Increase in activity ○ Decrease in activity

Meditator: I settle down for meditation in a relaxed, upright position, and slowly relax my body, from the toes of my feet to the top of my head. At the same time I focus my attention on an image of the Virgin Mary. Thoughts appear, interrupting and demanding attention. However, I just return to the image and these thoughts disappear. I try to build up the image in my mind, seeing every single detail, and slowly it becomes clearer and it becomes easy to hold the image in the stillness. The Virgin Mary is fixed in my mind.

Neuroscientist: At this point, there is an increase in activity in the attention association area and a

decrease in activity in the surrounding areas of the
frontal cortex. This is accompanied by an increase in
activity in both the visual and orientation association
areas. These areas are necessary to fix an object in the
mind. As activity in the parietal lobe decreases, the
individual's ability to see the boundary between 'self'
and 'object' also decreases. This accounts for the sense
of absorption into the object.

This increased activity in the visual association area of the
occipital lobe (Step 1, see illustration opposite) acts to stimulate
the hippocampus, again conveying the sense of the emotional
significance of the experience, as well as imprinting the
experience in the long-term memory (Step 2). The hippocampus
in turn stimulates the amygdala, which strengthens the
emotional significance of the experience (Step 3).

The amygdala also activates the body's autonomic response
through the midbrain (Step 4). First the relaxation system is
maximally activated, giving rise to a sense of blissful peace, and
then the arousal system is switched on, giving rise to a sense of
alert clarity of mind.

If the visual object or image that has been used as a focus of
attention has a special spiritual or emotional significance, the
emotional content of the image acts to exaggerate this chain of
events. Accordingly, the autonomic nervous system's response
to a 'meaningful' image, such as a statue of the Virgin Mary or
of Buddha, is greater than that seen with a neutral image, such
as a candle flame.

Eventually, as in passive meditation, this process in the right
hemisphere travels to the left hemisphere, leading to a greater
loss of sense of space and/or time, a loss of the ability to
comprehend the experience in rational terms, and an inability
to describe the experience using language.

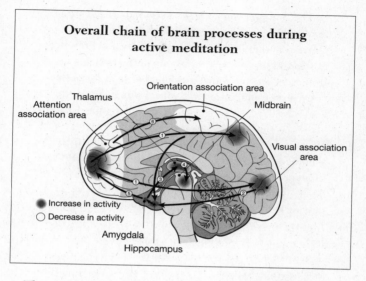

Overall chain of brain processes during active meditation

The continuous visual fixation on a particular object, combined with the gradual dissolving of the self/non-self boundary (left parietal lobe) and the expanded sense of connectedness and unity (right parietal lobe), can result in the meditator feeling absorbed by or becoming one with the object. As activity in both the left and right parietal lobes is dampened, awareness of the boundary between object and observer becomes blurred (step 5).

Meditator: After a while, the boundary between myself and the image slowly seems to dissolve. It's a strange feeling; losing my sense of self and being in the embrace of something boundless, infinite. I'm absorbed into the image; the distinction between it and me becomes blurred, irrelevant. A wave of blissful peace washes me from head to foot in light and love (later I interpret this to be an overwhelming sense of the Virgin Mary's love for us all). As I emerge from meditation, I'm often struck by a clarity and presence of mind I don't associate with my normal waking life. In these moments sometimes images, vivid memories, or even profound insights, surface, along with strong emotions.

Neuroscientist: At this point, the decrease in activity in the orientation association area in the parietal lobe produces an autonomic nervous system response. As before, this accounts for the wave of peaceful bliss passing through the individual's body. The autonomic nervous system also triggers the clarity of mind; however, as it is accompanied by a dampening of activity in both the right and left parietal lobes, there is a corresponding decreased awareness of 'self' during this meditative experience, so the individual is unlikely to become aware of this clarity until after the meditative session.

Important neurobiological features of meditation

The two schemes that Newberg and d'Aquili mapped out to explain the processes of passive and active meditation have been criticized by some other researchers. Undoubtedly, they are a gross simplification of what occurs in the brain during meditation. Other researchers have noted different changes in the brain during meditation, and certainly the exact changes will differ considerably between individuals and depending on the particular contemplative technique being practised. Since the definition of meditation is so broad, neurobiological studies of the meditating brain have included a wide range of different techniques, so it is perhaps not surprising that the findings of these studies are also so diverse. However, despite this simplification and variations in experience, Newberg and d'Aquili's schemes do reveal a number of crucial elements of the meditative experience.

Intention and attention

First, and most importantly, meditation calls on intention and attention. Intention is the conscious will to behave in a particular way and attention is the conscious focusing of awareness in order to perform the intention. In other words, when we intend to meditate, we refocus our attention to allow us to do so.

Attention and awareness can be considered to be equivalent: one is aware of what one is attentive of. Sensory awareness or attention arises from activity in a reverberating neural loop in the brain between the prefrontal areas of the cortex and the thalamus, the gateway through which all sensory information must pass. Neurones in the thalamus are thought to focus the attention, in much the same way as we might use a searchlight on a dark night. As we turn the beam of light on one object or another, we become aware of each of these objects in turn. If attention is combined with intention, this scan of the environment might prompt us to walk over and examine one of the objects.

When we're sitting at rest, our attention is usually unfocused. Accordingly, our mind is filled with random, and possibly not very relevant, thoughts and sensory information. These make us even more inattentive, as we are distracted by one thought after another.

In meditation, the intention to sit down and meditate triggers the thalamus to refocus attention either inwards or outwards, depending on the particular contemplative technique chosen. This brings the mind to one-pointed or all-embracing focus. As both the amount of incoming sensory information and the number of distracting thoughts are reduced, a meditator's attention becomes stronger and stronger, until attention can be effortlessly sustained. A crucial element in this process is the detachment from thoughts. A meditator is aware of thoughts as they arise but does not analyse them.

Modification of sense of self

As we have seen, meditation brings about a change in brain function that modifies the sense of self. This can partly be understood in terms of a shift from left-brained (ego-centred) to right-brained (holistic) thinking. Attention is primarily a right-brained function. In the Western world, most individuals navigate through their everyday life in a fashion dominated by left-brained thinking. Attention therefore precipitates a shift towards right-brained thinking and in doing so allows an 'escape' from the confines of an ego-centred existence, a possibility of expanding awareness beyond mere self and the immediate environment.

However, the right hemisphere is also involved in generating a sense of the three-dimensional space we occupy, so loss of a sense of self cannot be purely derived from a switch to right-brained activity. Newberg and d'Aquili suggest that the changes in the sense of self during meditation can also be understood in terms of a decrease in activity in the region of the brain known to be associated with the construction of our self/non-self boundary, in both the left and right hemispheres. Our awareness of where self ends and non-self begins is compromised as activity in the orientation association areas of the two parietal cortices is dampened.

Brain wave states during meditation and mystical experiences

Activity levels in the brain can be gauged by the electrical frequency at which different regions are operating. An electroencephalogram (EEG), which records the combined electrical activity of large groups of neurones within the brain, can allow us to see and measure this activity, and provides a way of monitoring brain changes in different stages of meditation or mystical experiences.

Brain wave rhythms

A number of dominant brain wave frequencies can be normally identified on a typical EEG recording. These different frequencies are also referred to as brain wave rhythms or states.

Beta waves

Beta waves have a high frequency. They represent small groups of neurones flickering into and out of action across the brain. These bursts of activity have been associated with higher cognitive functions, such as active thought, attention or problem-solving, and they are associated with waking consciousness: the eyes are open and the brain is involved in processing incoming sensory information.

Alpha waves

Alpha waves have a slightly lower frequency. They represent large groups of neurones that remain active for a longer period of time than those involved in the generation of a typical thought. They are associated with a state of relaxed wakefulness: the eyes might be closed and the mind and body relaxed, in a state of inner-directed attention and non-linear, holistic thinking.

Theta waves

Theta waves are even lower frequency and have been associated with the creative subconscious mind. They represent activity of large groups of neurones that gives rise to the ability to form mental images, and thus imagination. Accordingly, they are particularly prominent in both dreaming sleep and deep meditation. They are also the predominant brain wave rhythm in children between the ages of two and five years old. At no point in our lives are we more creative and imaginative than in our early childhood. Children can create entire worlds in their minds, escaping into this fantasy landscape when awake and asleep. This is the consequence of strong theta wave activity.

Delta waves

Delta waves are very low frequency and are associated with unconscious mind. They are so small that they are often obscured by electrical impulses resulting from movement of the large muscles in the neck and jaw. They reflect activity in the large array of neurones of the unconscious that give rise to intuitive thought, empathy and instinctual insight. Delta waves occur mainly in infants, sleeping adults or in adults with brain tumours or those in a coma.

Other brain waves

Recent evidence supports the existence of a number of additional, distinct brain wave states, including gamma, hypergamma and epsilon. **Gamma waves** are fast beta waves, associated with the ability to assemble a number of small fragments of information stemming from different regions of the brain, into a single, coordinated picture. **Hypergamma waves** are extremely high-frequency waves that, together with the very low-frequency **epsilon waves**, are thought to be associated with high-level cognitive abilities, such as insight, psychic powers and out-of-body experiences.

Correlating brain wave states and stages of meditation

Typical EEG recordings are very complex, and their interpretation requires considerable training. As a result, it is hard to extract any hard and fast rules about EEG changes evoked by meditation. The variety of different techniques used in meditation is reflected in a highly variable EEG signature, making it impossible to give one EEG signature that is relevant for all types of meditation, or all types of mystical experiences, or in all individuals. However, there are some generalizations that can be made. Broadly speaking, a meditative experience is characterized by a less complex EEG recording over time. This is supposed to be the result of the 'switching off' of irrelevant networks.[2] A meditative experience is also often associated with changes in alpha, theta and gamma brain waves.

Changes in alpha waves

In the early stages of meditation, there is an increase in alpha wave activity. This is associated with a state of relaxed wakefulness. There is also a greater synchronization of alpha wave activity across the two hemispheres. Behavioural studies have shown that greater synchrony between the two hemispheres is linked with both external attention and a lack of expectations.

The increase in alpha waves peaks during a transcendental experience, should one occur. (Transcendence is defined as a state of being or existence above and beyond the limits of material experience.) The amplitude, or range, of alpha waves is much larger during a transcendental experience than at any other point in a meditative session. These EEG changes have been shown to persist once the meditative experience has ended, and therefore the awareness obtained through the meditative experience spills over into both the normal waking and sleeping states.

Changes in theta waves

In the small number of meditators who report transcendental experiences while meditating under laboratory conditions, this 'peak', 'absolute unitary being', or 'kensho' experience is associated with an increase in theta activity. In one study of yoga practitioners, the intensity of the bliss experienced by long-term meditators was found to correlate positively with the level of theta waves in the frontal cortex – the more prominent the theta waves, the more blissful the experience. There was also a correlation between the level of frontal theta waves and intrusive thoughts – the more prominent the theta waves, the lower the incidence of thoughts.

The increase in frontal theta waves is larger in the left hemisphere than in the right.[3] Since it is known that the left hemisphere is associated with positive emotions and the right hemisphere with negative emotions, the disproportionate increase in theta waves in the left frontal cortex explains why meditation, on the whole, is associated with positive emotions such as bliss.

The increase in theta waves also accounts for the highly imaginative images that can punctuate meditative sessions. Similar findings have been reported in an early study of transcendental meditation.[4]

Changes in gamma waves

In deeper stages of meditation, some meditators display high-frequency beta or gamma waves, which are also thought to reflect a subjective state of transcendence. Gamma brain waves are associated with the assembly of numerous different fragments of information into a single, coordinated picture. Antoine Lutz and his colleagues carried out a study of eight, long-term Buddhist meditators who practised unconditional loving-kindness meditation, in which the point of focus is a particular person or group of people. During this type of meditation, not only did the recorded amount of high-frequency gamma waves increase, but gamma wave activity was also seen to become synchronized across the entire brain. An experienced meditator reported that four different meditative techniques were associated with four different subjective states of mind. These four independent states were all reflected as gamma wave activity, but the precise neurones involved in this gamma wave activity differed depending of the meditative technique used.[5]

Differences between novice and experienced meditators

Whilst investigating EEG readings during meditation at the Russian Academy of Medical Sciences, neuroscientists Aftanas and Golocheikine noted some important subjective differences between short-term and long-term meditators.[6] Subjects were split into two different groups. In the first group, the subjects had been practising meditation for less than six months, while those in the second group had been meditating for between three and seven years.

At a subjective level, the short-term meditators reported feeling more uneasy and restless than the long-term meditators, while the long-term meditators felt more bliss and were less likely to be interrupted by thoughts than the short-term meditators.

There were also differences in the EEG changes observed in the two groups. The long-term meditators showed increases in theta and alpha waves in the frontal cortex, whereas short-term meditators showed no significant changes. The long-term meditators also showed an increased synchronization of alpha waves, which is a reflection of the decrease in external awareness and a lack of expectations. Short-term meditators, on the other hand, exhibited a desynchronization of alpha waves, which is thought to reflect the impact that their expectations have on their meditative practice. This offers a neural clue as to why novice meditators are advised not to meditate with expectations, as these very expectations can act as an obstacle to their progress. It would appear that expectations elicit a desynchronization of the alpha content of the two brain hemispheres, thwarting attempts to withdraw attention from the external environment.

Meditation embraces a wide range of different techniques and this inherent variability may, at least partially, explain the variability seen in the EEG findings of different studies. It is clear that the precise changes in the EEG recording can differ depending on the contemplative technique used. Of the two types of meditation described above, transcendental meditation involves one-pointed focus on a particular object, image or mantra, whereas loving-kindness meditation involves the generation of a particular state of mind (one of compassion) without particular attention directed towards any one object. This fundamental difference between these two techniques appears to be reflected in different EEG signatures.

The relaxation response

The process of meditation is thought to be superimposed over a general relaxation response mediated by the parasympathetic nervous system, which includes the following:

- decrease in oxygen consumption
- reduction in the elimination of carbon dioxide
- reduction in heart rate, respiratory rate, blood pressure, lactate levels in the blood (indicative of a reduction in muscle activity) and muscle tone
- increase in blood flow to the internal organs
- increase in the temperature of the fingers
- increase in skin resistance, as measured using the GSR.

There are four different elements that facilitate this relaxation response during meditative practice:

1 a mental device to shift a meditator's mind from logical and externally orientated, left-brained thinking to intuitive and internally orientated, right-brained thinking.
2 a passive attitude; meditators should not attempt to rate

their technique, or become frustrated at distracting thoughts, since these processes cause anxiety and therefore counteract the relaxation response.

3 a comfortable position so that minimal muscular effort is necessary to maintain an upright position, without becoming too comfortable and therefore promoting sleep rather than relaxation.

4 a quiet environment, especially in the case of novice meditators, to minimize any noises that may cause distraction or anxiety.

Brain wave patterns and levels of consciousness

In 1978 Maxwell Cade, the British psychophysiologist and biophysicist, published research centred on the correlation between brain waves and transcendental experiences. Using a specialized EEG, he found that different levels of consciousness, and indeed different psychological stages of meditative practice, could be correlated with different brain wave patterns. Recordings from the brains of experienced meditators, irrespective of their background training, revealed the same overall pattern of brain waves during meditation. These different elements are accompanied by changes in the EEG trace consistent with decreased activity in the sympathetic (arousal) nervous system and increased activity in the parasympathetic (relaxation) nervous system. The subjective and objective responses of the different stages of relaxation are summarized below (adapted from *The Awakened Mind* by Maxwell Cade and Nona Coxhead[7]).

Cade believed that the brain wave state achieved through regular meditative practice was one of a number of possible brain wave patterns adopted by the human brain. There were in fact, he said, five observable levels of consciousness: dreaming sleep, the hypnogogic and hypnopompic states (see pages 105–107), the waking state the meditative state and the Awakened Mind.

Relaxation response (% change)	EEG rhythms	Subjective description of state of relaxation
Below 25%	Intermittent beta and alpha	Begin to relax; may report difficulty in stilling the mind and/or restlessness
25–35%	Reduced beta and continuous alpha	May report feeling dizzy or having a fuzzy consciousness, as well as the frequent desire to fill the mind with everyday thoughts
35–45%	No beta; continuous alpha, and intermittent theta	Sense of calmness and relaxation; intermittent attention; flashback memories of childhood or the past
45–55%	Continuous alpha of ever lower frequencies, with almost continuous theta	Sensations of floating, lightness, rocking and swaying. Attention becomes sustained and imagery clearer and more frequent
60–70%	Continuous alpha and theta of ever lower frequencies	Extremely vivid awareness of breathing, heartbeat or other bodily sensations; effortless attention; fluctuation between external and internal awareness

70–80%	Continuous theta and alpha of a frequency close to the alpha/theta border	Lucid impression of altered state of consciousness that is deeply satisfying and associated with intense alertness, calmness and detachment
Over 80%	Very little electrical activity except occasional delta	Synthesis of opposites into a higher unity (see page 134), combined with a new intuitive insight into old problems

First level of consciousness

This is experienced during dreaming sleep. At this first level we have no awareness of the external world and are focused on the internal world of the imagination, on the dreams arising from the personal (and collective) unconscious. In some cases, we can become aware of the fact that we are dreaming, a state referred to as lucid dreaming. This state is particularly interesting, since an awareness of the imaginary nature of the dream content allows us to manipulate the content of the dream, for example to resolve issues of conflict buried in the unconscious, as well as being able to access memories and images normally suppressed during waking consciousness. Dreaming sleep is associated with a predominance of the lowest frequency brain waves: delta waves.

Second level of consciousness

This includes both the hypnogogic (between waking and sleeping) and hypnopompic (between sleeping and waking) states. In these states, we either start to rouse from sleep, or just begin to drop off to sleep, when we realize that we are unable to move. In most cases, this paralysis is accompanied by the feeling of a 'presence', often malevolent, threatening or evil, that appears to monitor events just out of sight. Often this feeling is reinforced by the presence of auditory, visual and tactile hallucinations, as well as by out-of-body experiences. At this level we have an increased awareness of the external world, but have yet to regain full control over our bodies and still have access to the imagery prevalent in the dreaming sleep state. This level is associated with a predominance of the slightly higher frequency brain waves: alpha waves.

Third level of consciousness

This is our everyday level of consciousness. In this state, we are aware of our external environment, and this awareness leaves us in a state of permanent arousal. As this level of consciousness is mediated by the conscious mind, it is accompanied by the thoughts and higher cognitive functions that stem from this conscious mind. Accordingly, the brain wave patterns contain predominantly beta waves. However, since many of us spend most of our day in a left-brain mode, there is an imbalance in the beta wave content of the two hemispheres, with more beta content on the left than the right.

Fourth level of consciousness

This is the state of the meditating mind. We withdraw awareness from the external environment and focus attention on the internal environment (self-awareness) or on a mental visual image. As we have already seen, the meditative state is associated with an absence of the beta waves that dominate waking consciousness, and an increase in both alpha and theta waves in the brain wave pattern. The meditative state (and higher states of consciousness) can therefore be differentiated from waking consciousness (and lower states of consciousness) by the presence of multiple frequency bands (alpha and theta) rather than just a single band.

Fifth level of consciousness
This is a state of lucid awareness. In this higher waking state, we maintain both continuous self-awareness and continuous awareness of the external environment. This heightened state of awareness is characterized by a brain wave pattern that has been referred to by Cade and his colleagues as the Awakened Mind. The wave pattern contains similar amounts of alpha and theta waves to that seen in the meditative brain state, but also includes the beta waves that are inevitably associated with higher cognitive functions. Unlike the third level of consciousness or waking state, however, the beta wave content is balanced over the two hemispheres, indicating a balance between left-brained thinking and right-brained feeling.

A spectrum of consciousness

The link between different levels of consciousness and distinct brain wave patterns has implications. First, and perhaps most profoundly, these observations suggest there is a continuous spectrum of consciousness, reflected in a continuous spectrum of different brain wave patterns. This confirms the link between mind and matter, although it cannot answer the age-old question: which came first, consciousness or the brain?

Secondly, although meditation has proved to be an effective and controlled method of training meditators to access higher levels of consciousness, often these involve many years of trial and error. Recordings of the electrical brain wave pattern allow meditators to learn about their internal brain wave states, and the effects that these different states have on their behaviour. Furthermore, EEG technology can be used to train meditators to switch between different brain wave states to optimize brain performance. This exciting field of development is explored in greater depth in Chapter 5.

We have seen how, irrespective of whether meditation is passive or active, whether a meditator uses a mantra or an image of religious significance or even a set of contemplative movements, the ultimate result is a shift from left-brain to right-brain function and dissolution of the 'self', leading to an awareness that our everyday reality is only a small fraction of the awareness of which the brain is potentially capable. Not only do these findings provide a rational, scientific explanation for the subjective experiences of meditation, but they also suggest that our brains have been designed so that an appropriate key, such as that provided by meditation, can unlock this hidden potential. It also appears to blur the boundaries between science and spirituality, challenging our traditional view of the relationship between these two disciplines. This is what we shall explore in the next chapter.

Bridging science and spirituality

THE SEPARATION OF SCIENTIFIC AND SPIRITUAL MATTERS HAS ITS ORIGINS IN THE EARLY STAGES OF THE DEVELOPMENT OF HUMAN THOUGHT. In the sixth century BC, Greek philosophers proposed the existence of a divine being on a higher plane of reality, and therefore separate from humanity. From this belief sprung the division of spirit and matter that has dominated Western thinking for many centuries. However, it now appears clear that our scientific investigations should inform our spiritual investigations, and vice versa. From this a new scientific field has been born: neurotheology. Not only does this chapter highlight the importance of ritual, myth and the indispensable 'leap of faith', but it broaches the inevitable question: will neurotheology eventually prove or disprove the existence of God?

Shifts in perspective and belief

For many centuries, the accepted thinking in the Western world was that reality could be given a single, complete and unambiguous description in human language. Initially, it was assumed that religion would provide this all-embracing description of reality. However, in the 16th century, scientific investigations undermined one of the so-called 'religious truths', that the Sun and other planets revolved around the Earth. Copernicus, the founder of modern astronomy, presented a theory claiming that the Earth actually rotated around the Sun. Although Copernicus was to die before he fully realized the impact of his work and the controversy it would cause, two of his ardent supporters, the Italian astronomers Galileo Galilei and Giordano Bruno, would suffer the consequences of upholding a theory that defied the view of the Church. Charged with heresy, Galileo was sentenced to life imprisonment and Bruno was burnt at the stake.

This marked the beginning of the scientific revolution through which the limited religious worldview, then rigidly based on the literal interpretation of the scriptures, was slowly replaced by a scientific worldview based on experimental observations. People then looked to science for the much-sought all-embracing description of reality.

It is only now, after 400 years or so, that this scientific worldview is also slowly being viewed as limited. In Chapter 2 we saw how investigations at a quantum level challenged the presumed objectivity of science. We are now forced to face the fact that neither the scientific nor the religious worldviews are capable, single-handedly, of providing a definitive and complete description of reality. A survey of the religious beliefs of American scientists, conducted in 1997, found that 40 per cent believed in a personal God.[1] Furthermore, many of the pioneers in quantum physics in the early 20th century – Einstein, Schrödinger, Heisenberg, Bohr, Eddington, Pauli, de Broglie, Jeans, Planck, Bohm – have been described as mystics. It should therefore be possible to reconcile these different disciplines.

In 1934, Einstein wrote: 'You will hardly find one among the profounder sort of scientific minds without a religious feeling of his own.'[2] However, this opinion is not universally accepted. There are still many scientists who view religion as a superstitious relic from the past, and maintain that science will eventually provide a complete description of reality. Similarly, there are still many religious thinkers, particularly within the more fundamental sects of the major world religions, who claim that science cannot reveal 'Truth', going so far as to consider scientific discoveries as blasphemous if they contradict the literal interpretation of scripture.

Religion and spirituality

When examining the relationship between science and religion, it is important to make the distinction between religion and spirituality. Spirituality is based on direct, personal experiences of non-ordinary dimensions of reality or higher states of consciousness, and does not require a specific place or mediator for communication with the Divine. Religion, on other hand, although originating in the majority of cases from spiritual experiences of prophets, saints or ordinary followers, has largely lost touch with these spiritual sources, becoming instead organizations comprised of specific places of worship, defined rituals and official mediators.

Throughout the history of most major World religions, the emphasis has not been placed on the propagation of personal spiritual experiences, but on the pursuit of power, control, politics and possessions. Many people, particularly in the Western world, have therefore become disillusioned with religion. This does not, however, automatically equate to disillusionment with spirituality. While our postmodern society has seen a decline in the number of people attending church every week, it has also seen a growth in spirituality and its slow disentanglement from the rituals and dogma of organized religion.

The extraction of the underlying core spirituality from the exterior shell of the dogmatic and ritualistic traditions of religion holds the key to its reconciliation with science. The much-publicized similarities between the insights of the new physics and those inherent in the Perennial Philosophy suggest that our understanding of both scientific and spiritual matters will increase as a result of interdisciplinary exploration. The esoteric traditions are all based on the personal experiences of the mystics and, as we will see, these subjective mystical experiences appear to be, at least at some levels, comparable to the scientific observations that form the basis of our scientific tradition.

Both our scientific and spiritual experiences are limited by our involvement in the process of observation, and so we cannot expect to obtain or devise a complete description of reality from any one approach alone: we can only piece together the limited glimpses of reality obtained through these different approaches. Despite obvious differences, both types of experience can be seen to provide valid insights into the nature of our reality. The integration of scientific and spiritual knowledge is therefore essential to our quest for a more complete description of reality, and indeed a more complete understanding of meditation and mystical experiences.

The Perennial Philosophy

In 1945 Aldous Huxley published a book entitled *The Perennial Philosophy*, in which he proposed that common elements could be found in the traditions and lore of primitive cultures worldwide, as well as in all the major world religions. While these religions have considerable superficial differences, the esoteric teachings at their core have striking similarities. Huxley referred to this core of common teaching as the Perennial Philosophy.

There are several main tenets of the Perennial Philosophy that relate to what this chapter has to say about the nature of mystical experience:

- There is an Ultimate Reality that is both universally immanent in creation and transcendent to it. The limited reality that we can apprehend with the five physical senses is embedded within, and sustained by, a limitless Ultimate Reality.
- This Ultimate Reality cannot be reached or described using the rational mind; it is therefore inherently incomprehensible and paradoxical.
- There is something in the deeper eternal 'self' of a human being, distinct from the personal ego, which is similar to, or even identical to this Ultimate Reality.
- This Ultimate Reality is the ground of all being,

through which we are all interconnected.
- Through spiritual and moral practice an individual can experience awareness of and achieve union with this transpersonal reality.
- Once an individual has become aware of a connection with this Ultimate Reality, this awareness is then accompanied by a growth in compassion and wisdom.

Neurotheology: reconciling science and spirituality

As both science and spirituality are increasingly seen as complementary aspects of a greater whole, each capturing a differing and partial representation of a greater reality, a greater understanding is thought to stem from a greater integration between the two disciplines. As a result, a new discipline has arisen that appears to blur the edges between them: neurotheology. Researchers in this field are attempting to unravel the structures and processes in the brain that mediate mystical experiences. Not only does this research provide valuable clues about our ability as humans to have mystical or transcendental experiences, but it also reveals the role of particular brain structures and aspects of mind in the processes that give rise to belief and/or experience of a transcendent reality, equated with the Divine.

Contrasting scientific and mystical experiences

Conventionally, in Western society, scientific and mystical experiences are considered to be as different as chalk and cheese. Scientific observations are thought of as an objective examination of the nature of reality; mystical experiences as a subjective examination of the nature of reality. However, not only has the objectivity of science been called into question through quantum observations (see Chapter 2, page 68), but it is also becoming increasingly clear that there are some striking similarities between the scientific and mystical approaches.

Both approaches involve an interpretation of reality, one mediated by the use of mathematics and the other by the use of words and symbols. Indeed, the parallels between the mathematical formulae used by scientists and the words and symbols used by the mystics is one that is often missed. All represent something else; they do not describe reality directly, they only symbolize certain aspects of reality. All language, including that of mathematics, can therefore at best offer an incomplete description, a limitation acknowledged by the founders of quantum theory. As a result, they have been unable to describe the quantum reality in terms of a precise set of mathematical laws, in much the same way as mystics have always stated that experiences of the ultimate reality are indescribable and cannot be accurately represented using words or symbols. Both the scientific and mystical approaches are inherently limited by the tools at their disposal – as Wittgenstein observed: 'The limits of my language mean the limits of my world.'[3]

Similarities . . .

The personal experiences of the mystics, which form the basis of all mystical traditions, can also, in many respects, be likened to the experimental observations that form the basis of scientific discovery. In scientific experiments, the experimental apparatus that enables the investigation can be thought of in terms of a filter, through which we view only a partial representation of reality. By choosing the apparatus, we choose a filter and in doing so select the aspect of reality that is to be examined. If we want to measure the electrical activity of the heart, we use an electrocardiogram (ECG) scan. If we want to see the internal anatomy of the heart, we use a computerized tomography (CT) scan. At the quantum level, as we saw in

Chapter 2, the choice of the apparatus determines whether sub-atomic particles exhibit particle-like properties or wave-like properties. Heisenberg summarized this as 'what we observe is not nature itself, but nature exposed to our method of questioning'.

Similarly, in mystical experiences, our brain acts as the apparatus or mediator of our investigations of reality. Whereas our right brain captures and stores a complete image of the experience, our left brain filters the experience in line with our previous conditioning. It is well established that we all construct a mental and emotional framework, as a result of conditioning, through which we interpret our experiences. This conceptual map can be seen to be the equivalent of a filter placed between ourselves and the experience, and it gives rise to the subjective nature of personal experience.

Since both the scientific and mystical methods of enquiry employ filters that focus and limit our investigations of reality, neither can be expected to provide a complete representation of reality. Just as quantum theory has shown us that the observed behaviour of matter is only a partial representation of its true nature, the observed behaviour of a single brain cell is only a partial representation of its true nature and the glimpse of reality we perceive through mystical experiences is only a partial representation of its true nature.

. . . and differences

There are, however, a number of important differences between the methods employed by scientists and mystics.

First, the reality examined by a mystic is profoundly different from that examined by a quantum physicist. In mystical traditions, through the use of meditation, the mystics directly, but subjectively, examine the Ultimate Reality or the 'ground of being'. In modern physics, through the use of mathematics, physicists indirectly, but objectively (although, as we have already seen, the objectivity of science has been called into question), examine reality at a quantum level. Through quantum theory we have discovered that matter is not the 'ground of being' – it is seen instead as a condensation of an underlying, highly interconnected field or reality. Some scientists, most notably David Bohm, have equated this underlying field to the Ultimate Reality. However, while practitioners of the mystical traditions claim to access and experience the Ultimate Reality, the field underlying and giving rise to the quantum reality cannot be experienced; its existence is merely inferred from the experimental evidence. This again exposes the limitations of the scientific method.

Secondly, mystical experiences are portrayed as highly personal; no one description of a transcendental experience is presumed to portray accurately and definitively all transcendental experiences. They are seen as different facets of a single diamond, if you wish. Mystical experiences can also be evoked under many different conditions, and have been reported to occur spontaneously. Insight is therefore gained from the identification of meaningful patterns in these experiences rather than the precise nature of these experiences or the means employed to 'achieve' them. A single scientific observation, on the other hand, is often assumed to be a definitive description of an experience that remains accurate only for as long as the experimental conditions are kept constant. In fact, one of the fundamental principles of scientific research is that of repeatability: the same observation must be made on x number of occasions for it to be considered valid. Since it seems unlikely that any two mystical experiences would be exactly the same, and, in a single individual, no set of 'experimental' conditions can be said to elicit a uniform, let alone an identical, mystical experience, it is not realistic to expect that the observation will be made on more than one occasion. This lack of repeatability is also one of the major obstacles facing parapsychological research, as the psychologist Susan Blackmore noted: 'the spontaneous phenomena, which form an important part of parapsychology, are not and cannot be expected to be repeatable at will.'[4]

Thirdly, although mystical experiences can be intellectually analysed or conceptualized to a degree, ultimately they transcend intellectual experience, theories and insights. Descriptions of the underlying quantum field may sound intellectually similar to descriptions of the mystical Ultimate Reality, but these scientific models remain theoretical concepts, products of the rational mind.

Depths and breadths of the human mind

Historically, our study of the human mind has been undermined by the subjective nature of our investigations. The expectations of both subject and experimenter can influence the outcome of a so-called objective experiment, and this loss of objectivity has had profound implications on the value placed on so-called subjective evidence. It is becoming increasingly clear that the accumulated knowledge gained from subjective experiences can offer a valuable insight into the workings of the human mind, and therefore the human brain in states such as meditation.

Many researchers believe that the mind and brain interact in a complementary manner. The brain is seen to be a collection of neural elements and processes that give rise to a holistic function: the mind. The mind as a whole can be viewed as emerging from the whole brain and distributed throughout it.

The Swiss psychologist Carl Jung, proposed, like Freud before him, that the mind was split into different levels or strata. This is usually explained in terms of an iceberg (see opposite). The conscious mind – our conscious thoughts, perceptions, emotions and memories – represents the tip of the iceberg. Below the surface is the personal unconscious. This is a repository for all our repressed thoughts, perceptions, emotions and memories. Not only are our most traumatic memories stored in the personal unconscious, but also

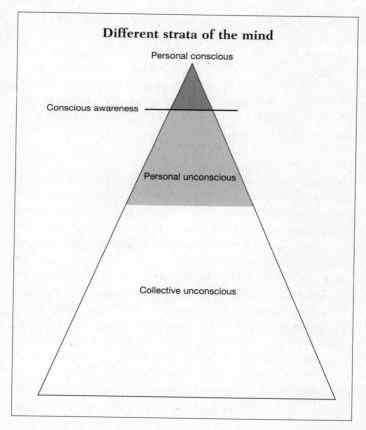

Different strata of the mind

Personal conscious

Conscious awareness ———————

Personal unconscious

Collective unconscious

emotions that should have been expressed at a certain point, but weren't for whatever reason, and all the aspects of our nature that we have yet to acknowledge as part of ourselves (our shadow side). Jung also described a deeper part of the unconscious, which he referred to as the collective unconscious: the repository of the entire human experience. In the collective unconscious, information is stored as a series of templates or motifs, which Jung referred to as archetypes. We shall be looking at the role of archetypes in myth-making later in this chapter. Although Jung was unclear of the origin of these archetypes, suggesting both that they were pre-determined patterns genetically transferred from one generation to the next through evolution and that they were images stored in a consciousness created by the interconnected activity of billions of individuals over time. In both descriptions, however, the collective unconscious can be viewed as a collection of templates that define and structure human experience.

The cognitive operators

The research of Andrew Newberg and Eugene d'Aquili describes the human mind as being composed of seven processing units, called cognitive operators. Each of these represents a specific function of the mind. Collectively, these cognitive operators allow us to interpret and assign emotional significance to our experiences.

The holistic operator produces the 'big picture', the ability to see how objects or experiences fit into a wider context. It allows us both to grasp the architectural theme of a building, and to experience the sensation of unity or oneness with all things in creation.

The reductionist operator allows us to dissect problems or objects into their component pieces, and analyse these components in a rational, logical and sequential fashion. It is undoubtedly responsible for the reductionist approach that has dominated scientific discovery in the West, and it is also the likely source of the widespread fragmented perception of the world.

The causal operator identifies the cause of a particular experience or series of events, and is responsible, at least in part, for our drive to understand why things happen. The causal operator not only gives rise to the drive for scientific discovery, but has also given rise to countless different religions and ideologies, each of which purports to know the ultimate cause of all things.

The abstractive operator generates abstract concepts, such as object categories (such as: an orange and a pineapple are both fruit). It allows us to suggest a way in which two pieces of information are related, assembling these pieces of information into a complete theory that explains their relationship. It gives rise to scientific theories, ideologies and religious or moral beliefs.

The binary operator allows us to make sense of our world by dividing it into pairs of opposing concepts, such as light–dark, good–evil and inside–outside. In the mental map formed by the binary operator, these concepts are seen to be separate and antagonistic; they act as anchors that aid our comprehension of the world around us, but can also be the source of conflict if in a state of imbalance.

The quantitative operator, as the name would suggest, allows us to define objects in terms of their quantity (there are *three* oranges). The ability to measure time and distance are crucial to our survival in this physical reality, and it has also given rise to most of our scientific theory.

The emotional value operator assigns emotional significance to a particular event. Whereas the other cognitive operators allow us to structure and interpret our physical reality, the emotional value operator imbues this interpretation with an emotional response. It gives rise, through the actions of the limbic system as a whole, to the emotions evoked by our experiences, and therefore determines our physiological response to these experiences.

Dissecting a bowl of fruit

As incoming sensory information is processed in the cerebral cortex, it is examined by one or more of the cognitive operators.

When we see a bowl of fruit:

- the holistic operator 'sees' a bowl of fruit;
- the reductionist operator 'sees' a collection of different pieces of fruit;
- the causal operator suggests that the bowl of fruit has been placed in view so that the fruit is eaten;
- the abstractive operator identifies the presence of oranges, apples and bananas;
- the binary operator notices that some fruit has fallen outside the fruit bowl;
- the quantitative operator counts the number of different types of fruits;
- and the emotional value operator makes you feel happy at the sight of the bowl and the thought of sampling something from it.

All the cognitive operators are thought to be mediated by distinct neural networks found in the parietal cortex, with the exception of the emotional value operator, which is located within the limbic system, most of which lies deep within the temporal lobe. Most of the cognitive operators can be found in the left hemisphere, and therefore give rise to the aspects of human intellect associated with left-brained activity. These cognitive operators work together with other regions of the cerebral cortex involved in producing thoughts, maintaining attention and facilitating language skills.

There are examples of people who lack one or more of these

cognitive operators. People with autism, for example, do not appear to have a fully functional abstractive operator, as they are often incapable of abstract thought. Most of us, however, possess all seven of these cognitive operators, and there is evidence that at least some of the operators are pre-programmed in the circuitry of the developing brain: humans are hard-wired to perform these particular cognitive functions that allow us to experience the world, and to interpret these experiences.

An ever-changing personal worldview

Evolutionarily speaking, the cognitive operators within our brain have evolved to allow us to evaluate potential threats in our environment, thereby promoting our survival. Other members of the animal kingdom respond to immediate threats: a bird flies away when it detects a cat in its immediate environment. Humans, however, can perceive *potential* threats, real or imagined, as well as immediate threats. Although this ability undoubtedly promotes survival, and accounts for our extraordinary success as a species, it also leads to a constant state of arousal (or fear).

What is the outcome of all of this processing? The cognitive operators extract meaning from the incoming sensory information and construct a mental map of the scene. This map is used to guide our immediate behavioural response as the cognitive operators trigger responses through the autonomic nervous system: the sympathetic nervous system springs into action raising the heart rate, blood pressure and breathing rate to ready us for any necessary evasive action, and slowing down unhelpful functions, such as digestion, appetite and the need for sleep. The cognitive operators also activate the amygdala, which modulates these responses, leading to a state of heightened mental alertness. In the short term, then, the cognitive operators provide the means to respond to incoming sensory information, whether it is to run away from danger or walk over to a bowl of fruit and choose something to eat.

The consequences of this behaviour generate additional incoming sensory information. This is again processed by the cognitive operators and our mental map is altered accordingly. If, for example, the chosen piece of fruit was rotten, this information would alter the mental map of the scene. The tempting bowl has become a bowl of potentially rotten fruit, the mental map is refined in line with this extra information, and our behavioural response is modified accordingly.

This never-ending cycle is the process through which we continuously fine-tune our behaviour in line with our experiences, encoding this learning in our mental map of a particular scene. This mental map in turn forms part of our entire conceptual map determining our perception of the world and our actions within it. Every experience leads to the formation of new mental maps or the modification of existing mental maps. Every experience therefore leads to subtle changes in our personal overall view of the world.

There is also a wealth of anecdotal evidence indicating that a single experience can completely transform a person's personal worldview. In these cases, the experience is so profound such that it alone, without any external validation, can lead to a permanent shift in perspective; a permanent refocusing of the psychological lens, so to speak. Examples of such transformative experiences include severe trauma and significant life events (marriage, parenthood, divorce, bereavement, for example), and also mystical experiences and near-death experiences.

Modifying behaviour

The process through which our worldview, and therefore our behaviour, is modified clearly involves structures throughout the brain. Sensory information is channelled into the relevant brain regions. This information is then processed by the cognitive operators that lie deep within the parietal lobe, and assigned emotional value through the diverse brain structures that comprise the limbic system. This processing triggers a response, which requires activity in areas of the frontal cortex, associated with motor function and higher cognitive skills such as planning, attention and intention. Our worldview therefore emerges from the functioning of the entire brain.

The ability of the brain, and therefore human behaviour, to modify itself in this manner is the secret of our evolutionary success, both as individuals and as a species. Individually, we modify our behaviour according to experience in an attempt to improve our chances of survival.

We have seen how simple modes of behaviour can be encoded by subtle changes in the strength of the connections between neighbouring neurones (see page 65). Here we are seeing how the psychological filter that dictates both our perception of the world and our behaviour within it is modified in line with our experience. Together they provide invaluable learning at two levels: in terms of changes in the functioning of single neurones or small circuits of neurones, and secondly, in terms of the changes in the overall functioning of large groups of neurones distributed throughout the brain.

The neuropsychology of mystical experience

Nature may abhor a vacuum, but we humans abhor chaos. Not the messiness of an untidy bedroom, but random senselessness. We yearn for order, to see sense in the world, and we seek to discern it or impose it wherever we can.

Pattern-seeking animals

The psychologist Stuart Vyse and his colleague Ruth Heltzer conducted an experiment in which participants were asked to navigate through a virtual maze using the directional keys of a keyboard. Participants in the first group were awarded points for reaching the lower right-hand section of the maze, while those in a second group were awarded points randomly. Both

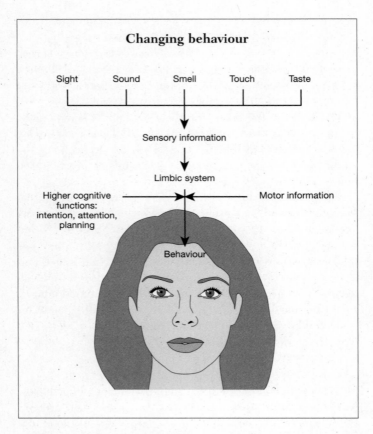

Changing behaviour

Sight   Sound   Smell   Touch   Taste

Sensory information

Limbic system

Higher cognitive functions: intention, attention, planning   Motor information

Behaviour

groups were then asked to explain how they thought the points had been awarded. Both groups identified patterns in how the scores had been assigned. The participants in the second group detected a rationale even though the scoring had in fact been random, because they expected to see a pattern in the scoring, for it to have a meaning.[5]

Humans have evolved as pattern-seeking animals because patterns are abundant throughout nature: in the spiral patterns formed by the galaxies of our universe, the shells of snails and the seeds in a sunflower's head; in the intricate designs of snowflakes and the hexagonal symmetry of honeycombs.

The study of complexity, otherwise known as chaos theory, provides insights into how these patterns develop in nature. Chaos theory maintains that the behaviour of living systems is essentially random and therefore unpredictable. However, when the behaviour of living systems is monitored over time, order often arises spontaneously from disorder. This spontaneous process – self-organization – gives rise to the patterns that we can see throughout nature as order arises from chaos.

There are numerous examples in nature of apparently random irregularities that, when examined closely, can be seen to display a surprising level of order: regularity within irregularity. This regularity can even be mathematically modelled and the resulting images from computer simulations, referred to as fractals, are now well known. These fractal patterns can be seen to underlie apparently random shapes such as coastlines, mountains and clouds. Although we undoubtedly do occasionally see patterns where there are none, on an everyday basis we use our pattern-seeking ability to our survival advantage.

Some staunch materialists, such as Professor Richard Dawkins, suggest that our perception of the Divine is yet another pattern created by our minds to confer meaning to a chaotic existence. This pattern is passed down the generations, from parent to child, and passed from one individual to another through spiritual practices. This view is correct in so far as our *images* of God, as the benevolent old man or a vengeful judge, are indeed acquired through social and cultural conditioning, and passed from parents to their children. However, the *experience* of God is something that is indescribable and abstract. The experience cannot be passed from one individual to another; it can only be experienced.

Fractal pattern generated by mathematical model

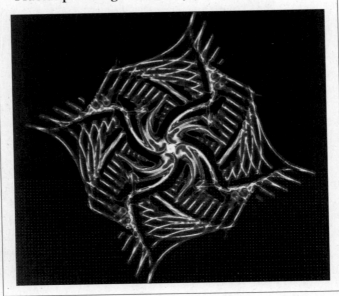

Religion as a process of self-maintenance

A fundamental component of all our myths is the existence of a higher reality or being that transcends our physical reality. The answers to our existential questions are seen to transcend rational analysis or comprehension, and therefore do not stem from the cognitive processes that we traditionally associate with analytical, left-brained thinking. Instead, these answers are seen to stem from something that lies beyond human understanding; something that is superhuman.

The form that this superhuman being takes is dependent on the religious context in which these existential questions are posed. Authoritarian religions, such as John Calvin's Protestantism, Jehovah's Witnesses, Christian fundamentalist sects and Mormonism, emphasize sorrow, guilt and sin. God is a personal God, a judgemental God who sets down rules of obedience. Dogma, ritual and rules of conduct are paramount, and show similarities to authoritarian political regimes. Humanistic religions, such as early Buddhism, Taoism, Zen Buddhism and early Christianity, on the other hand, emphasize joy, peace and the importance of self-realization. God is a transcendent reality, an infinite source of energy that pervades all things. Faith is based on experience rather than dogma, and spiritual discipline stems from the individual rather than being imposed by an authoritative figure.

In all religions, the concept of God is an omnipotent entity that guides our behaviour. In authoritarian religions, this guidance is from an external source, whereas in humanistic religions it comes from within. Irrespective of the source, this guidance acts to modify behaviour in line with the moral framework the religion lays down. Newberg and d'Aquili describe religion as a 'method of self-maintenance'.

We have seen that one of the main functions of the mind, represented by the causal operator, is a drive to establish causality. We strive to find the underlying cause of every situation or series of events, and this inevitably leads us to question what ultimately caused our existence, and what will therefore ultimately determine what will happen to us when we die.

Many different religions and ideologies have sprung from this drive, and each provides an alternative code of conduct. In this way, many religions provide a ready-made worldview through which all adherents' experiences are filtered. This ready-made worldview was devised by an individual mystic, be it Jesus, Buddha or Mohammed, from their own personal spiritual experiences, using the countless archetypes or images stored in the collective unconscious. The mental map they provide, therefore, is similar in concept to the mental maps that we continuously construct to aid our navigation through this physical reality. The myths and rituals of different religions act as a way of teaching individuals this mental map, and then reinforcing this learning by frequent repetition of the fundamental concepts. This mental map then acts both to guide individual behaviour and to dictate perception of experiences.

Accounting for variations in religious experiences

After thousands of years of culturally dependent categorization, humankind has created a rigid conceptual map of certain aspects of our physical reality – for example: pigs cannot fly; humans do not have naturally green hair. Not only is this mental map formed by the cumulative experience of a person's lifetime, but it is also an imprint of all of our personal, societal and cultural conditioning. It defines our goals and expectations, it dictates the way in which we perceive the world, and it provides an explanation for our experiences. However, although this map acts as a framework through which we can communicate details of our experiences to others, it also limits our ability to describe and understand personal experiences.

Everyone's conceptual map is slightly different. This means that even an identical stimulus is likely to elicit totally different responses in different people – your description of the perfect sunset will no doubt differ from my own, and a piece of music that can reduce me to tears may have little effect on you.

This subjectivity accounts for the diversity seen in the accounts of mystical or religious experiences. The importance of an individual's conceptual map in determining the material content of a mystical experience cannot be downplayed. As a result of conditioning in our childhood, most of us possess a fairly rigid image of the Divine. Conventional Christianity and Judaism have propagated an anthropomorphic image of a personal God who can be both benevolent and vengeful. In Hinduism, on the other hand, there are many different deities, each a physical manifestation of one of the many facets of the

Divine, while in Buddhism, the Divine is pictured as an transcendent reality that gives rise to and pervades all things.

These images determine our relationship with the Divine and dictate the forms of our experience. In a Christian, a mystical experience may be interpreted as the appearance of the Virgin Mary, whereas in a Hindu it may be interpreted as a vision of Ganesha, the elephant-headed God. Joseph Campbell, an American writer on mythology and comparative religion, suggests that, although images and stories of spiritual significance may vary from culture to culture, 'they remain valid as metaphors which express our experience of something beyond the human'.[6]

Understanding of the role of ritual

Ritual behaviour can be defined as a structure or pattern of behaviour involving two or more individuals and relating to a common goal or purpose. It is usually highly repetitive and rhythmic in nature, and acts to integrate or synchronize behaviour within each of the participants and within the group as a whole.

Ritual can also be performed by an individual during meditation. Whereas in communal ritual, there is communication between individuals, in private ritual, communication is between the meditator and Ultimate Reality or the Divine. Examples of ritual behaviour abound in both ancient civilizations and our modern world: the spinning meditation of the whirling dervishes or a Catholic's praying the rosary.

It has been known for more than 50 years that a repetitive stimulus, such as a flashing light or a chanted tone, is capable of synchronizing the activity of the brain, tuning the overall brain activity to the frequency of the incoming stimulus. This offers a possible process through which repetitive, ritualized behaviour, such as chanting or whirling, can produce changes in the brain wave states that we saw on pages 97–102.

Since ancient times, humanity has used ritualistic behaviour to tap into hyper-arousal and hyper-relaxation states. Rapid ritualistic behaviour, such as long-distance running or prolonged periods of dancing, activates the sympathetic nervous system and gives rise to a sense of heightened awareness and alertness. This is often also accompanied by the sense of both 'going with the flow' and being sustained by vast amounts of energy channelled in from an outside or deeper source. Slow ritualistic behaviour, such as chanting or prayer, on the other hand, activates the parasympathetic nervous system and gives rise to a sense of tranquillity and restful contentedness. This is accompanied by a blissful lack of thought and bodily sensations. In Chapter 3 we saw that both the parasympathetic and sympathetic nervous systems play an important role in the unfolding of mystical experience via techniques that both restrict and expand attention (see pages 84–94).

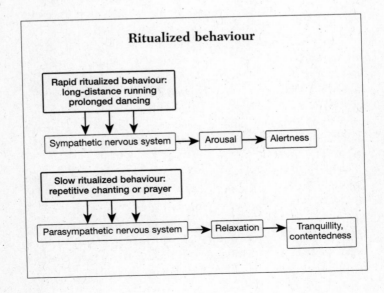

The importance of myth-making

Myths were devised, and continue to be devised, in an attempt to explain the inexplicable. Jung taught that myths were a creative expression of images or patterns stored in the collective unconscious (his 'archetypes'). Four of the most famous of Jung's archetypes are:

- the shadow (those aspects of our nature that are not expressed, but stored in the unconscious)
- the anima (the feminine aspect of a man's nature)
- the animus (the masculine aspect of a woman's nature)
- the Self (the image of the Divine).

These archetypes can be seen in the folklore of ancient civilizations, in the teachings of Jesus, Buddha and Mohammed, and in what Jung called 'the dreams, visions, and delusions of modern individuals entirely ignorant of all such traditions'.[7]

As images, these archetypes arise from the function of the right hemisphere. They are then interpreted by the left hemisphere. The archetype of a 'mother', for example, contains all the aspects of human nature that we attribute to motherhood: unconditional love, protection, nurture. This pattern allows us instantly to recognize maternal behaviour. An archetype can therefore be considered to be the right-brain equivalent of the mental maps that form in the left hemisphere. In the left brain, however, this overall impression is analysed and split into components, some of which are deemed to be desirable and others that are not. The different types of processing attributed to the two hemispheres account for the different format in which this information is stored: images for the holistic right brain and networks of associated concepts, or mental maps, in the rational left brain.

The creation of myths

Newberg and d'Aquili have proposed that all myths have a common framework. First, an existential question is posed: 'How was the universe created?'; 'What happens when we die?'; 'What is the purpose of life?'. Then, the issue is presented in terms of conflict between two apparently irreconcilable opposites, such as good–evil, life–death or love–hate. Finally, a possible resolution is presented, usually in terms of the reconciliation of polar opposites, such as unity from dualism. This cognitive trick can, when delivered in the context of ritual, confer a real sense of resolution of the presented question, and a sense of its existential implications. But what has happened in terms of brain function?

The first and second stages trigger activity in the left hemisphere. The comprehension of language, including such fine-tuned functions as the comparison of concepts and the naming or categorization of objects, are all left-brain activities. The third stage, however, triggers activity in the right hemisphere. The ability to comprehend unity and reconcile polar opposites requires right-brain processing. This switch between left- and right-brain thinking illustrates, once again, the cooperative behaviour of the two hemispheres. Activity in the left hemisphere drives activity in the right hemisphere; the quest for meaning to our experiences triggers a shift in brain function that allows us to perceive the 'big picture'.

So myths can be viewed as an ancient form of psychotherapy. They allow us to tackle an existential problem by tapping into the intuitive and holistic processing of the right brain. The solution the myth offers cannot be gained from rational and logical analysis; it arises from 'looking at the big picture' and, often, in the reconciliation of polar opposites. When presented in an appropriate manner, and in an appropriate context, the 'recipient' experiences the resolution of the problem in the myth at a deep, existential level, leading to a resolution of internal conflict. Since this resolution does not stem from the cognitive processes that we customarily work with, we attribute these answers to something that lies beyond human understanding. The concept of God is therefore implicit to our myths, and it represents the deeper source of knowledge that stems from right-brain activity. This knowledge is only perceived to originate from an outside source because it derives from a region of the brain that lies outside the domain of the 'ego'.

In the past, people have attributed this right-brain knowledge to wisdom communicated to them directly by God, or angels, or even aliens. All these concepts are products of the left brain, the storyteller, frantically trying to make sense of knowledge that it cannot claim ownership over. The humanistic religions all stress that true knowledge comes from within; we already have all the answers, we merely need to re-learn how to access them.

Personal myths and shared mythology

So far we have only examined myths in the context of their use in communities to provide resolution to life's existential unknowns, to derive meaning from the experiences of humanity as a whole. However, we also use myths at an individual level, to derive meaning from our own personal experiences.

Just like the storytellers of old, our left brain invents, using the wealth of images contained within the collective unconscious, elaborate stories that confer meaning to our experiences. Since these personal myths employ collective images, when recounted to others they often 'resonate' with these individuals; that is, their right brain identifies the 'truth' in what they are being told. The personal myth therefore becomes a communal myth. This forms the basis of all the mystical traditions: Jesus, Buddha and Mohammed, and their disciples after them, were all enigmatic storytellers. The commonality in the messages conveyed by these different mystical traditions, as identified in the Perennial Philosophy (see page 113), is likely to be a consequence of the common pool of images or archetypes drawn from the collective unconscious.

Thwarted attempts to prove or disprove the existence of God

When Vilayanur Ramachandran presented his findings on the link between temporal lobe epilepsy and religious experience (see page 77), scientists and the media alike almost immediately spread the word that a 'God module' had been found in the human brain. For some, Ramachandran's research promised to yield a method of eliciting mystical experiences in even the most fervent atheists; a fast-track to the experience of God. However, as we also saw in Chapter 3, Michael Persinger's following research demonstrated that an individual's ability to experience God is related to the sensitivity of their temporal lobe to change (temporal lability), and in doing so, he provided a possible explanation for why some people deny the existence of anything other than that contained in our physical reality. It is possible that the fervent scientific materialists of our time insist that mystical experiences, and indeed God, are unreal because they do not possess sufficient temporal lability to experience them. This does not mean that some individuals are innately incapable of having these experiences; it just means that these experiences do not come easily and are therefore unlikely to occur spontaneously and will be hard to evoke artificially using Persinger's helmet.

For others, this avenue of research yielded proof that God doesn't exist; God and all religious thought are artefacts of brain function. Take the following example. If the specific areas of the brain involved in the visual perception of an apple were to be stimulated, we would report, unsurprisingly, that we had seen an apple. The apple is not real; it is an artefact of brain function, conjured up by stimulating the appropriate neurones in the brain. Similarly, when Persinger experimentally stimulated the temporal lobe, many subjects reported having a religious or mystical experience. These experiences were also artefacts of brain function, created by the artificial stimulation of certain neurones in the brain. However, the possibility that, at some other time, one of those participants could experience a 'real' stimulus should not be excluded. Does the fact that a non-real apple can be induced call into question whether apples actually exist? No, of course not. When we see an actual apple, specific regions of our brain will light up with activity, capturing and interpreting this sensory information. The fact that these processes occur does not undermine the validity of the experience. We have still observed a 'real' apple. Similarly, we can identify specific brain regions and functions that are associated with mystical experiences; this does not undermine the validity of these experiences.

We all possess the neural circuitry to experience the full range of experiences; we are primed to receive these 'real' stimuli. Persinger's experiments showed that neural circuitry exists in the brain that is capable of mediating mystical experiences; in other words, his experiments revealed that humans are hard-wired to have mystical or religious experiences.

The inherent limitations of the scientific method, and indeed our interpretation of all experience, ensure that we will never be able to definitively prove or disprove the existence of God. All experience is subjective in nature, as it is filtered and interpreted by our own individualized conceptual map. It is wrongly assumed that an observation of a chemical reaction is more objective than an experience of a reality transcending our everyday reality. The fact that the experience of this transcending reality is processed in the right-brain presents a further obstacle to this quest. The cognitive operators usually at our disposal to categorize and interpret our experiences lie within the left brain and are therefore inactive during the experience. The experience of God therefore defies rational analysis and cannot therefore, in any way, be evaluated in a truly objective manner. As Persinger himself says: 'It's something that is invisible, non-physical, eternal, and everywhere. If you're going to define it that way, then there's no way you could ever test it.'[8]

The experience of God: product of a delusional mind?

Some members of the scientific community have gone as far as to suggest that religion, and therefore presumably mystical experiences, are a feature of delusional psychological processes. Religious fervour is seen to be a mental illness in the same way as, say, schizophrenia. There are undoubtedly examples of extreme religiosity that could be considered pathological, when individuals' beliefs cause them to do harm to themselves or to others. However, this does not mean that religious experiences per se are pathological; it is the individual's interpretation of them that can become pathological – in fact, the word 'delusion' can be defined as erroneous beliefs that usually arise from a misinterpretation of perception or experience.

Take the example of a patient with schizophrenia. It is well known that patients with schizophrenia can often misinterpret their experiences, giving them undue significance or distorted meaning, resulting in paranoia: a person innocently walking behind them on their way home becomes a CIA agent tracking their every movement. The experience of a person walking behind them is real; it is their interpretation of the experience that it is distorted. People with disorders of the self, such as patients with schizophrenia, interpret religious or mystical experiences differently from so-called 'normal people'. Mystical experiences often involve a widening of the sense of self and, in the absence of a fully functional and unfragmented ego, this broadened awareness can act to exacerbate a disorder.

In the majority of cases, people reporting a religious or spiritual experience, or expressing deep religious or spiritual beliefs, do not have any other indicators of an underlying mental health disorder. Mystical experiences are a common feature of human existence; they can be experienced by both 'normal' and 'abnormal' people; and they can arise both spontaneously and in response to the practice of internal and external techniques. To view mystical experiences themselves as pathological, rather than merely viewing their consequences in a small proportion of individuals as pathological, is 'throwing the baby out with the bathwater'. It also highlights our restricted perception of what is 'normal' and what is 'pathological' behaviour.

The indispensable 'leap of faith'

In some people, the shift from left-brain to right-brain activity during a mystical experience produces a sense of the experience originating from an external source – they may report hearing the voice of God or a vision of angels, for example. Because activity in the left hemisphere is dampened during these experiences, the ego, the 'self', is essentially switched off and the experience is felt in the right hemisphere, which lacks an internal storyteller to make sense of the experience. This can result in confusion about the experience, as knowledge stemming from within is perceived to originate from something distinct from self.

This inability to rationalize mystical experiences means a 'leap of faith' is required. This leap represents an acceptance of the knowledge gained through the experience, despite the fact that the information cannot be attributed to the ego or substantiated through rational analysis. It requires belief in its truth without proof.

We routinely distinguish between real events, arising from the external world, and imagined ones, emanating from our mind. When imagined experiences are captured and analysed by the left brain, they are attributed to fantasy or imagination, based on their sensory and emotional content, their familiarity, and our personal belief systems. During mystical experiences, when the sense of personal self is switched off, it becomes more difficult to attribute experiences either to the realm of 'real' or 'imagined'. The inability to analyse a right-brain experience rationally can confound this issue further, and it needs a leap of faith, expressing belief in the truth of the experience despite objections raised by the rational mind.

Mystical experiences are usually associated with an extremely strong sense of the 'truth' of the experience, that is perceived to stem not from our everyday reality, but from an extraordinary reality that both transcends and pervades the everyday. Our perception of this extraordinary reality is not constrained by our limited left-brain worldview. This reality is perceived as transcendent and pervasive, as something that extends beyond the 'self', but still encompasses it; something that cannot be intellectualized, but persists even when the mystical state has ended.

Our perception of everyday reality is constrained by our personal worldview stored in our left brain. However, mystical experiences bring awareness of the right-brain's worldview and slowly, over time, this more expansive worldview can become superimposed on the left-brain worldview, subtly changing our overall view, and behaviour.

The integration of knowledge gained from both scientific and spiritual investigations has given us a more complete picture of reality than either discipline alone could offer. We are hardwired to seek out both the cause and the meaning of our experiences, and it is therefore perhaps inevitable that many different religions and spiritual traditions have evolved, each claiming to provide the answers to how and why we exist. The myths and rituals associated with these different religious and spiritual traditions exploit innate functioning in the brain to expand awareness and promote a shift from left- to right-brained thinking. It is this shift that both defines the need for a 'leap of faith' and ensures that we will never be able to definitively prove or disprove the existence of God.

Hard-wired

A high-performance mind

OUR BRAIN CONTAINS 'HARD-WIRING' THAT ALLOWS US TO EXPERIENCE BOTH HIGHER STATES OF CONSCIOUSNESS AND AN ALL-PERVADING UNITY THAT CAN BE EQUATED TO GOD. It is perhaps not surprising, in our fast-paced Western society, that we have explored numerous methods of facilitating the attainment of these higher states. Meditation often involves many years of trial and error until a technique is refined, and there is no way of establishing whether an individual is practising the technique properly. Often, someone can practise meditation for months, even years, before realizing an error in practice. In this chapter we'll explore three strategies commonly used to enhance meditation and facilitate our attainment of altered/higher states of consciousness: psychoactive substances, biofeedback and neurofeedback. These strategies promise to increase our awareness of unconscious behaviour, optimize the performance of our brain and facilitate learning, memory and spiritual practice. It is therefore critical to be aware of the inherent benefits and risks of each of them, and to understand how they relate to meditative experiences.

Tripping into altered states of consciousness

Seeking to enhance our understanding of reality through the use of mind-expanding drugs did not begin with the psychedelic adventures of Western researchers in the Swinging Sixties. Psychoactive drugs have been part of the shamanistic practices of indigenous cultures in many parts of the world for centuries – members of the Native American Church, a religious movement originating among the Indians of the Southwestern USA, for example, regularly ingest peyote (a cactus containing mescaline) as part of their religious ceremonies. Even in Western society, a significant proportion of people the world over regularly use a psychedelic agent to elicit altered states of consciousness, and to tap into the creativity and intuition associated with these altered states.

In the late 1970s a group of researchers, unhappy with the

use of terms such as 'psychedelic' or 'hallucinogenic' to describe such a broad range of different substances with diverse effects, coined the term entheogenic. Originating from the Greek *entheos*, this term literally means 'becoming divine within'.

There are a number of entheogenic substances commonly used in a spiritual context to facilitate mystical experiences: psilocybin ('magic mushrooms'), LSD, ketamine, mescaline and DMT. Generally speaking, all these compounds, with the exception of ketamine, exert similar effects on the brain, so we shall look in detail at just psilocybin and LSD, which are widely used throughout the West, despite legal restrictions on their distribution and use, and ketamine, which can induce a state likened to a near-death experience.

Psilocybin

Mushrooms containing psilocybin have been used in a religious context by numerous ancient civilizations, including the Aztecs in Mexico, and continue to play a major role in tribal rituals in South and Central America.

Psilocybin was used in one of the most frequently cited experiments on entheogenic substances: the Good Friday Experiment, devised by Walter Pahnke in 1962. A group of students and faculty members at Boston University participated in an experiment in which they received, in random fashion, either psilocybin or an active placebo of nicotinic acid. They were then all monitored while they listened to a live broadcast of the Good Friday service in the basement of Boston University's Marsh chapel. Once the effects of the experimental drugs had subsided, the participants were asked to rate their experience in terms of nine mystical characteristics, including sense of unity, transcendence of space and time, paradoxicality and transiency. Pahnke reported that 80 per cent of subjects receiving psilocybin experienced at least seven of these nine mystical characteristics. Six months after the original experiment, most of those who had received psilocybin reported persisting positive changes in attitude and behaviour as a result of the experience.

Psilocybin, like many entheogenic substances, has a similar structure to serotonin. This is the neurotransmitter involved in the brain circuitry that modulates mood, sleep, appetite and emotion. Psilocybin binds to serotonin receptors throughout the brain, disrupting the neural circuits regulating the flow of sensory information through the brain. As a result, the frontal cortex becomes overloaded with sensory information. This explains the visual and sometimes multi-sensory hallucinations associated with psilocybin. Psilocybin also acts on serotonin receptors to change mood, producing, in many cases, euphoria. The activation of serotonin also regulates the activity of another neurotransmitter, dopamine. Decreases in dopamine levels lead to disturbances in thinking, illusions and impaired ego-functioning. The effects of psilocybin on ego-functioning is reflected in reports by some users of a feeling of dying or having died.

Exploring the long-term effects of the experience on Pahnke's 'guinea pigs', Rick Doblin found that, 25 years later, those who had received psilocybin still scored higher with respect to some of the mystical characteristics than those in the control group. They also reported that the experience had affected their lives in a positive way and that they remained grateful to have had the experience. Effects they described included an enhanced appreciation of nature and life in general, a deepened sense of joy, strengthened faith, reduced fear of death, greater acceptance of difficult life situations, and more compassion for people of differing cultures, belief and sex. Similar benefits were not reported from the control group.[1]

However, many of the participants who reported an experience with mystical content also said that there were elements that were frightening or disturbing. Some said they felt as if they were dying or had died, or that the experience was life-threatening. Others said they felt out of control and overwhelmed by guilt.

Of greater concern is the evidence that one participant had a psychotic episode in response to receiving psilocybin. This was not originally reported by Pahnke; presumably, he was concerned that this single episode would colour the overall findings of the study and undermine confidence in the controlled use of entheogenic substances. In people with a disorder that distorts the view of 'self', or a predisposition towards this type of disorder, psilocybin appears to produce a psychotic state that resembles schizophrenia. Its use is therefore not recommended where this might be a risk.

LSD

The hallucinogenic properties of LSD were accidentally discovered by the chemist Albert Hoffman in the early 1940s, and its experimental use was pioneered in the 1960s by the Harvard psychology professor Timothy Leary. Leary, who was soon dismissed from the faculty as a result of the nature of his research, formed the League of Spiritual Discovery and coined the phrase 'turn on, tune in, drop out'. Since then, LSD, also known as 'acid' or 'trips', has been used recreationally by countless people around the world.

As with psilocybin, the similarity of the LSD molecule to serotonin means that LSD binds to serotonin receptors throughout the brain. This triggers an increase in activity in the frontal cortex, resulting from an overload of sensory information, and a decrease in dopamine levels. Like other hallucinogenic substances, this causes detailed and spectacular sensory hallucinations. However, LSD is a far more potent hallucinogen than psilocybin, and the mixture of positive and negative elements experienced with psilocybin are amplified. Initially, people report visual distortions, the urge to laugh and slight dizziness. They can also experience feelings of euphoria and contentment. However, the experience can often become much more negative, as the effects reach their peak: a so-called 'bad trip'.

There is a considerable amount of variation in the LSD experience, both from one person to another and by the same person at different times. This can be explained in terms of both an individual's psychological state and the setting in which LSD is taken. Those who take LSD in a group are more likely to experience feelings of euphoria, whereas those who take LSD alone are more likely to experience anxiety, an inability to move and disruptions of speech. The use of LSD has also been associated with the development of psychoses in a small but significant proportion of users. The risk of psychosis appears to be higher in psychiatric patients[2] and a genetic link has been suggested to explain this increased susceptibility.

In the 1960s American psychiatrist Sidney Cohen conducted a survey into the safety of LSD and mescaline. Data from more than 5,000 people who had taken a total of 25,000 doses of one of these two drugs revealed that complications were 'surprisingly infrequent' and that, when taken under clinical supervision, these compounds were 'safe'.[3] Accordingly, for a while, LSD was used experimentally as a tool in analytical psychotherapy, particularly for the treatment of neuroses, phobias, obsessive-compulsive disorders, childhood schizophrenia, alcoholism and for the terminally ill. Later in his career, however, Cohen was to drastically alter his opinion on the safety of entheogenic substances. As more people used LSD in a recreational, rather than a clinical setting, reports of long-term psychotic effects, even suicides, increased. In response, Cohen published a statement in a journal sponsored by the American Medical Association (AMA) warning that the use of LSD could be associated with long-term psychosis and antisocial behaviour. This statement was the nail in the coffin for mainstream research into entheogenic substances.

Ketamine

Ketamine is a short-acting anaesthetic. It is most generally known for its use as a horse tranquillizer, but it is also used in surgery, particularly in developing countries and for children. Its use as an anaesthetic is linked to its ability to bring about a complete psychological dissociation from bodily sensations as well as, of course, its pain-relieving properties. At doses lower than those needed to anaesthetize, ketamine can also produce psychedelic experiences and hallucinations.

In a long-term Russian study, ketamine was integrated into a programme of psychotherapy for patients with alcoholism.[4] Patients described psychedelic experiences that had a number of common elements: a feeling of separateness or dissociation from the body, a loss of the sense of self or self-control, a sense of 'being dissolved and united with the universe' or an encounter with a higher being or God. The exact nature of the experience was determined by each patient's specific psychological state and no doubt, as is the case with mystical experiences, their spiritual and emotional states as well. About 75 per cent of the alcoholic patients maintained total abstinence from alcohol for more than a year following ketamine-enhanced psychotherapy, compared with only 25 per cent of patients treated conventionally. Patients also showed a positive transformation of their unconscious image of themselves and their emotional attitudes to themselves and other people, as well as an increase in spiritual development and positive changes in their values and purpose.

The ketamine-induced altered state has been likened to a near-death experience. During a so-called 'K-hole' experience, individuals often report features such as buzzing, ringing or whistling sounds, followed by the experience of travelling through a dark tunnel at high speed towards a light. The experience usually ends with the feeling of having died and being in the presence of God. These experiences are often accompanied by out-of-body experiences, similar to those reported by individuals during near-death experiences.

Ketamine binds to the receptors of a different neurotransmitter from psilocybin or LSD, the excitatory transmitter glutamate. By doing so, it dampens excitation within the brain. An explanation for this comes from New Zealand-born psychiatrist Karl Jansen's 'ketamine model', in which he suggests that ketamine mimics a naturally occurring substance in the brain that is released to protect neurones from the detrimental effects of such excitability. For example, in conditions such as epilepsy, hypoxia (low levels of oxygen in the body) and ischaemia (reduced blood flow through a part of the body), levels of glutamate in the brain can become so high that they cause the death of neurones in large portions of the cortex.

A near-death experience is suggested to be the subjective experience of this naturally occurring protective process; a psychological defence mechanism that alerts the person to impeding death and dissociates them from the intense emotions associated with death. The similar experience elicited by ketamine is proposed to be the result of its ability to dampen excitability in the brain, which results in the same psychological protective mechanism brought about by real brain trauma.

Ketamine can also produce paranoia, perceptual alterations, disorders of thought and cognitive deficits similar to those seen in patients with schizophrenia. These drug-induced episodes of psychosis have provided a valuable insight into the causes and possible treatment for patients with psychotic disorders, such as schizophrenia.

Entheogenic substances as a mystical technology

Investigations into the effects of entheogenic substances on the brain have provided further clues into altered states of consciousness. However, pharmacological routes to higher states of consciousness are fraught with difficulties.

The human brain is awash with different chemicals, the

levels of which need to be maintained within fairly narrow limits to ensure normal brain function. These same investigations have highlighted the role played by various neurotransmitters, such as serotonin, dopamine and glutamate, and disturbances in these neurotransmitter systems, in the pathophysiology of schizophrenia and other psychotic disorders. Drugs that threaten this carefully maintained balance therefore bring with them inherent risks of psychoses and other serious psychological implications. These risks are undoubtedly higher in those predisposed to these conditions, but there is also some evidence that people predisposed to psychiatric morbidity are also more likely than the general populace to experiment with entheogenic substances. In other words, those most at risk are the most likely to consume the drugs.

It remains incredibly difficult to unravel the specific roles that different neurotransmitters play in mystical experiences such as those produced through meditation. Schizophrenia has been suggested to be the result of imbalances in the dopamine and serotonin systems, but specific drugs that activate these systems have proved to be inadequate in its treatment. Similarly, it would be inappropriate to place an emphasis on a pharmacological manipulation of these systems to produce mystical experiences. Such a strategy, with its inherent risks, is unlikely to produce a long-term, viable alternative to methods that elicit mystical experiences through the global functioning of the brain.

Comparing drug-induced and natural mystical experiences

In Chapter 4 we saw that mystical experiences are characterized by a number of important features: they involve an expansion to unity awareness; they are ineffable; they are imbued with a conviction of 'truth'; they are transient; they are passive and so require no personal effort or will; they leave a sense that the 'self' is a limited construction of the mind; and they are usually accompanied by optimism, compassion and personal growth. Drug-induced experiences often exhibit just these same features and so, according to this definition, can be classified as mystical experiences. But are they quite the same, and can entheogenic substances provide a helpful route, even a more reliable, easier-to-achieve route, to higher states of consciousness than meditation?

Points of comparison

Overall, both in the short term and the long term, entheogenic substances can yield positive benefits; however, as is often the case with contemplative mystical experiences, drug-induced experiences are best accompanied by emotional support, spiritual guidance or even psychotherapy. We have seen that both the circumstances in which a substance is taken and the individual's state of mind can influence the nature and content of the experience. The same can be said for mystical experiences – both drug-induced and naturally induced experiences are highly variable in content and are heavily influenced by the individual's psychological, emotional, physical and spiritual state.

Timothy Leary emphasized that psychedelic experiences are influenced by 'set and setting'. Whereas Leary, for example, took his first dose of LSD in his forties, many present-day entheogenic experimenters are considerably younger, doing their first 'trip' in their adolescence or early twenties. Undoubtedly, the recent increase in the number of people reporting a long-term psychotic reaction to one or more of these drugs is partially related to their widespread, unsupervised use. As entheogenic substances have become more readily available on the 'black market', the shift from the clinical or spiritual use to their predominantly recreational use will undoubtedly have an impact on the effects that people, especially the young, experience.

Experiences elicited by entheogenic substances can have both positive and negative elements, and can, in a small number of people, lead to psychotic episodes that resemble schizophrenia. In clinical terms, expanding the mind appears to be contraindicated in patients genetically predisposed to psychotic disorders or with pre-existing mental health disorders. In these patients, psychotherapy should be used to address a person's emotional and psychological state before any benefit can be derived from the expansion of awareness that both meditation and entheogenic substances offer.

Shades of difference in experience

Meditation can be seen to elicit a gradual expansion of awareness, from a restricted egocentric view of ourselves into a wider sense of ourselves that acknowledges our connection with others and our environment. The integration of this growing awareness into everyday life requires emotional, psychological and spiritual changes. As the expansion of awareness that meditation brings about is a gradual process, a person's emotional, psychological and spiritual aspects can adapt in line with this growing awareness in a similarly gradual manner. With entheogenic substances, however, unless they are taken in circumstances similarly conducive to emotional, psychological and spiritual growth, the expansion of awareness can be hard to integrate into everyday life. The 'truth' revealed appears to be in conflict with values and beliefs already held and, without supervision, drug-induced mystical experiences rarely elicit the transformation in character and outlook associated with meditation.

In his book *Rational Mysticism*, John Horgan relates an experience he had after taking a 'supercharged' form of LSD. After an initial period of fantastic multi-sensory hallucinations, Horgan became overwhelmed with loneliness and the impossibility of his existence. This experience had a profound influence. It left him believing that all creation was a manifestation of God's personality disorder, that creation sprung from God's loneliness. Even many decades later he appears to be still searching for knowledge to help him reconcile his experience with his everyday existence.

Undoubtedly there is a risk of becoming attached or obsessed by the state of awareness conferred by a mystical experience, and all traditions emphasize that the achievement of any altered state of awareness is not the goal of contemplative practice. Often, entheogenic drugs are taken specifically to achieve altered states of consciousness, and the mystical experience becomes the goal of taking the drug. True meditation is not goal-orientated, and the development of an attachment to the experience will eventually block the experience. The risk of unhealthy obsession with obtaining particular states of consciousness is much higher with the use of entheogenic substances, particularly if they are used uncontrolled and unsupervised.

In some people, experiencing a transcendent reality via an entheogenic drug acts as a 'teaser' and encourages them to seek out other methods of accessing this transcendent reality without the use of external agents. In others, however, such a 'teaser' encourages them to use drugs on a regular basis, to give more frequent access to this transcendent state.

In controlled conditions and with the right support, entheogenic drugs can provide a glimpse into a transcendent reality. However, drug takers are reliant on these substances; they are not trained to reach these altered states of consciousness, they merely 'sit back and wait'. Contemplative techniques practised on a regular basis (at least daily) slowly train us to access altered states of consciousness and allows us to integrate the knowledge gained through this access into our everyday lives. It is not advisable to take entheogenic substances on a daily basis, and therefore people using these substances still lack a technique to expand their awareness with this degree of frequency.

It appears clear that pharmacological routes to altered states of consciousness do not represent a viable long-term alternative to meditation. Most entheogenic drugs remain controlled substances, and this reflects the inherent risks associated with their unrestricted and unsupervised use.

Biofeedback: shifting the threshold of conscious behaviour

Contemplative techniques have been refined over many thousands of years and, when practised properly, they efficiently elicit a series of processes within the brain that produce altered states of consciousness – experienced meditators can access these altered states of consciousness at will and with apparent ease. However, reaching this level of experience often involves many years of trial and error, and no one method works for all. This explains both the diversity of meditative techniques and the number of people who try several techniques before settling on the one that suits them. Even then, some people can practise a particular technique for many years without having any mystical experiences or significant insights.

This has led some researchers to believe that meditation is an inefficient technique that needs to be refined to make it more effective. This view may well be partially the result of the attitude of our fast-paced society: 'we want solutions, and we want them now'. However, on examining personal accounts of the contemplative path, it also seems clear that many people make slow progress because they fail to receive appropriate feedback on their technique or appropriate guidance when choosing a particular technique. In John Horgan's *Rational Mysticism*, the renowned transpersonal psychologist, Ken Wilber, stated that he believed that it was possible to 'dramatically increase people's actual chance of interior transformation' by matching each individual with an appropriate and effective mystical technique. This, he claimed, could be achieved by the use of innovative mystical technologies.

We have seen that the connections in our brain are very malleable, and that this provides the key to breaking out of conditioning. Much of our learned behaviour is encoded in brain circuitry that does not require any conscious involvement. By becoming consciously aware of unconscious behaviour, we can regain full control of it. Physiological reflexes, such as urination, can be brought under conscious control, and just as one can regain conscious control over this physiological reflex, psychological and physiological levels of arousal and relaxation can also be consciously modified. By becoming aware of the impact of different thoughts or environments on arousal, we can regain the ability to control that arousal, both in terms of state of mind and state of bodily relaxation. Leading psychotherapists and other scientists support the use of a GSR meter to increase awareness of unconsciously driven behaviour.

What is the GSR?

Galvanic skin response (GSR), also referred to as electrodermal activity (EDA), or more generally as biofeedback, is a psychophysiological response that can be measured in skin that contains sweat glands, such as the tips of the fingers. Typically, the GSR is measured via one or more electrodes attached to a person's fingers. The reading can be translated into a comprehensible format in a number of ways: as a moving trace on a computer monitor, as an auditory tone, or as changing frequencies of a flashing light or indeed the colour of that light. It can also be combined with computer gaming technology.

What the meter records is a measure of the resistance of the skin to the flow of a small electrical current. Skin resistance has been observed to change according to level of arousal: a high level of arousal is indicated by a fall in skin resistance (and a rise in the GSR reading), whereas a low level of arousal (or,

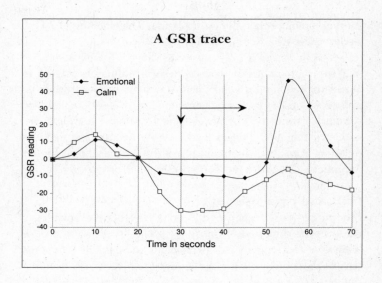

A GSR trace

to put it another way, a high degree of relaxation) is reflected in an increase in skin resistance and a drop in the GSR reading. The GSR is therefore widely considered to be a peripheral indicator of arousal in the brain. Its effectiveness in detecting emotionally charged issues has also led to its widespread use as a lie detector.

The sample GSR trace (see above) shows two readings, one taken from an emotional subject and the other from a calm subject. Initially, the two traces are fairly similar, rising and falling with the breath. However, on the presentation of a stimulus (indicated by the arrows on the trace), both respond with an increase in the GSR reading and thus an increased level of arousal. The response of the emotional subject is much more pronounced than that of the calm subject, and lasts for a longer period of time.

The neural processes involved in GSR remain undefined, although undoubtedly the sympathetic nervous system, with its 'fight or flight' responses, is involved. We have seen that this triggers, among other things, sweating. The opening of sweat glands leads to a dramatic fall in skin resistance, and this is reflected in rise in a GSR meter reading. We have also seen that the sympathetic nervous system is intimately related to both emotional and psychological states. Emotional outbursts, of both joy and anger, are accompanied by spikes in the arousal system, accompanied by physiological changes, such as heart rate, sweating and changes in skin resistance. A GSR meter is therefore capable of monitoring a person's arousal in response to particular circumstances or specific lines of questioning.

GSR and psychotherapy

Monitoring skin response during psychotherapy provides a guide to the direction of questioning – by revealing topics or memories that trigger an unconscious increase in emotional tension, it can identify issues that need to be carefully explored further. It can also be used to distinguish between psychopathic and 'normal' patients. Psychopaths have been reported to have a significantly smaller GSR change, and thus arousal response, to a conditioned stimulus, such as a small electric shock. The use of biofeedback in psychotherapy is still fervently supported by some of the psychotherapy profession, perhaps most notably by the psychotherapist Peter Shepherd, who has written profusely on the topic of biofeedback monitoring in this context, as well as devising a course that describes its possible uses.

Biofeedback and the anterior cingulate cortex

The effects of biofeedback have recently been explained in terms of its effects on the anterior cingulate cortex (ACC). One of the many structures in the limbic system, the ACC is a small structure tucked in between the brain's two hemispheres.

The ACC is involved in decision-making, and produces our emotional reaction to our decisions. When we are confronted with a choice between a number of behavioural responses to a

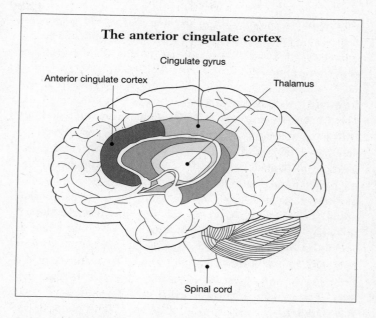

The anterior cingulate cortex

particular stimulus, the ACC evaluates how well things are going after we make a particular decision, and perhaps even before we become aware of the outcome of that decision. It is perhaps not surprising then that when a person is consciously monitoring their behaviour and its effects via biofeedback, that the ACC becomes activated.

The ACC is also intimately related to the sympathetic arousal system, and is believed to play a role in *intentional* changes in the level of arousal. Consider the following scenario. As you step off the curb into the road, your ACC determines the consequences of this action. If, rather foolishly, you have stepped into the path of a taxi, the ACC works with the sympathetic nervous system to bring about the response that propels you back on to the pavement. Similar increases in ACC activity can also be detected during meditation. Although the conscious rating of meditative practice is discouraged, it is only human to monitor our behaviour and its consequences. The ACC therefore observes the effect that meditation has on your level of arousal/relaxation, and judges whether you are performing optimally.

Learning to control arousal levels

The primary interest in the GSR meter has revolved around its ability to increase awareness in our own unconscious, automated behaviour. When we walk down the street, for example, in all likelihood our body will be a state of arousal. We are poised to take action should circumstances demand it, and our heart rate, breathing, blood pressure and muscular tension are all increased to prepare for this potential future action. A moderate level of arousal when walking down the street is necessary to avoid potential hazards, but many people are in a heightened state of arousal at all times, even at inappropriate times. As we saw in Chapter 1 (see pages 26–28), this constant state of arousal gives rise to elevated stress levels, and can lead to a wide range of stress-related diseases. A GSR meter draws attention to otherwise unconscious physiological processes by displaying information about them.

By seeing the effects of different stimuli – whether as a trace on a computer screen, the movements on a needle on a dial, a change in an auditory tone or a flashing light – we can become consciously aware of our physiological responses and train ourselves to modify our level of arousal and bring these responses under voluntary control. We can learn to use the biofeedback cues both to keep our arousal at a steady level and to change our arousal level at will. If we know, for example, that an increase in an auditory tone corresponds to an increase in arousal, we can make a conscious effort to keep the tone at or below a pre-defined level. Control over the biofeedback meter translates into control over the physiological process the meter is measuring. With time, we gain the ability to control the physiological process without the need for the visual cue provided by the instrumentation.[5]

In the Introduction, I described the first time I had a training session with a GSR meter and my excitement at seeing a visualization of my internal state of arousal. On a physical level, the GSR meter enabled me to watch how different breathing and visualization techniques could lower the level of arousal in my body, and so release tension. It allowed me to see the impact of shrugging my shoulders and releasing the tension in them and how by sequentially going through my body, relaxing different muscles, it was possible to lower the GSR reading.

On a mental level, the GSR also allowed me to watch how certain thought patterns, or interrupting stimuli, could raise the levels of tension in my body. The sound of a banging door caused a short-lived peak in my GSR reading, whereas the sound of my phone ringing caused a rise in my GSR reading that lasted for several minutes as I mentally fretted about who the call may have been from.

We are often not aware of how much tension we store in our body (and mind). Like a spring, we absorb the energy of events during the day, compressing ourselves tighter and tighter, storing more and more tension. Some people, aware of this building tension, strive to release it in activities such as a game of squash or a brisk walk. Others, however, store this tension up, giving rise to a plethora of stress-related disorders and a tendency towards uncontrollable releases of tension, such as emotional outbursts and drug taking.

The insights gained from GSR training leak into everyday life. In encouraging a greater awareness of my body and the areas in which I normally store tension, the GSR training made it more natural for me to be aware of these problem areas at intervals throughout the day and release the building tension with a stretch or a short walk. It also made me more aware of how certain thought patterns could increase tension in my body, and the benefit that relaxing my body had on these thoughts and indeed on my frame of mind.

Using GSR to enhance meditation

The GSR can also be used to facilitate the search for the meditative technique that best suits you. By watching the effect of different techniques on a GSR meter, you compare how each works on your arousal and relaxation levels, and can choose a technique that works most efficiently for you.

Relaxation, of course, plays an important role in the preparatory stages of meditation. The majority of traditions teach that meditation should begin with a gradual and progressive relaxation of the muscles of the body, together with an unforced and progressive quietening of the mind. By recording your level of arousal during a meditation session, you can see, afterwards, how changes to your level of arousal correlate with your subjective experiences. It can help you learn how different stages of relaxation feel.

A GSR recording can also be useful to a guide or spiritual counsellor offering additional feedback on the relative 'success' of different techniques. It allows a mentor to check that a meditator's subjective experiences match the information provided by the GSR recording. Someone who relaxes their body but is engaged in daydreaming, for example, may think that he is meditating; however, the GSR recording will reveal an arousal response to these daydreams. In being able to visualize the meditator's internal state, a guide or counsellor can obtain an insight into the individual's subjective state, and offer advice on any necessary 'tweaking' of technique.

Neurofeedback and the high-performance mind

Biofeedback has also been used in the context of modifying, even possibly optimizing, brain function. This technique, referred to as neurofeedback, is a promising new therapeutic option for patients with behavioural disorders such as ADHD, and affective disorders such as anxiety and depression, but it also bears much promise in improving relaxation techniques and meditation.

As we saw in Chapter 3, our brain waves vary in frequency with our state of consciousness. Put briefly, beta waves dominate active thought, such as when we are problem-solving or concentrating; alpha waves are indicative of relaxed wakefulness and non-linear, holistic thinking; theta waves are associated with creativity, imagination and visual imagery; while delta waves predominate when we are sleeping.

Neurofeedback allows us to visualize these variations as they are triggered by different stimuli, both external and internal, and under different physical conditions. By seeing our brain wave pattern translated into a visual image, we can become consciously aware of the way in which external experiences are reflected in our internal brain state. Furthermore, we can become much more aware of how our internal state influences our cognitive abilities, our performance in the physical world and our general sense of wellbeing.

What is brain entrainment?

In Chapter 4 we saw that a repetitive stimulus, such as a flashing light or sound, can tune brain activity to a particular frequency; the frequency of the incoming stimulus (see page 132). This ability has been used for many centuries in rituals such as chanting and whirling, but it is now also harnessed by visual–audio entrainment technology systems. In these systems, now becoming widely available, patterns of light stimuli, administered through special goggles, are combined with patterns of auditory stimuli, administered through headphones, to gently retune (or entrain) brain waves to frequencies that are more conducive to, for example, concentration or relaxation.

If this is difficult to envisage, picture this: imagine striking a tuning fork designed to produce a frequency of 440 Hz. If this oscillating tuning fork is moved near to another 440 Hz tuning fork, the second fork will also start to oscillate, and at the same frequency. The first tuning fork is said to have entrained the second fork.

Not only does this technology entrain the frequency of the brain wave to the frequency of the incoming stimuli, but the presence of a repetitive visual or auditory input also acts to focus the attention.

Benefiting from neurofeedback

There is a growing body of mostly anecdotal evidence suggesting that brain entrainment can be used to elicit changes in an individual's level of consciousness: that waves of delta and theta frequencies may enhance creativity, improve sleep and reduce anxiety, while waves with beta frequencies may improve attention and memory, and waves with alpha frequencies may produce a sense of relaxation. Alpha/theta feedback has been reported to improve artistry in musical students and dance performance in ballroom and Latin dance champions. Such feedback has also been associated with personality changes, such as being more compassionate, emotionally stable, socially bold, relaxed and satisfied, all of which are at least partially related to improvements in mood.

Undoubtedly, the exact effects of various neurofeedback wave protocols will differ considerably from one individual to the next; however, in general, it does offer a method through which we might become more aware of the influence that different brain wave states have on our everyday performance.

The concept of a high-performance mind was put forward by Anna Wise, who had previously worked closely with Maxwell Cade, who formulated the states of consciousness given on pages 103–107. She proposes that it is possible to master the ability to adopt different internal states at will, and therefore access the level of consciousness that is most appropriate and beneficial to a particular situation or task at hand. She describes a number of different psychological techniques that can be used, in conjunction with neurofeedback technology, to develop an awareness of our different internal states and learn to access different levels of consciousness, facilitating processes that are innate to the functioning of the human brain. Over time, neurofeedback can train us to become aware of different internal states, giving us the ability to consciously switch between these different states. After a certain, and variable, amount of training, there is no longer any need to rely on the neurofeedback apparatus, as we gain the conscious ability to switch between states.

Imagine that you could, at will, switch into the optimal beta wave activity when your boss presents you with a report to write on a tight deadline, or switch into the optimal alpha wave activity when you want to relax at the end of the day, or even switch into the optimal alpha–theta wave activity when practising a musical instrument or sketching a drawing. Most of us do this to some degree, but we are rarely consciously aware of our internal state – there are enough distractions demanding our attention – so we rarely switch between different brain wave states in the most efficient manner.

If you can train yourself, using neurofeedback, to elicit on demand the beta wave states associated with attention and memory, you become more adept at triggering these states in everyday life; you become more attentive and productive. Neurofeedback has been reported to improve attention, and cognitive performance in general, in healthy volunteers.[6] Improved attention improves performance, whatever the task happens to be. When your attention is all over the place, the amount of focused attention devoted to each of the individual tasks at hand is small. As a consequence, some tasks will be completed, but probably in a rushed and error-prone manner, while others will not be done at all. If, on the other hand, each task is given your full attention, with no more than a fleeting thought of other tasks, in the end, all tasks will be completed.

This switch in the way in which your attention is distributed can be achieved by the conscious intention to focus the attention on a particular task. Neurofeedback does not provide this switch; it merely demonstrates, unequivocally, that you can have conscious control over your internal state and, indeed, the focus of your attention by exploiting an innate switch function of your brain.

Attention is a crucial trigger of the brain processes involved in meditation. All the contemplative traditions emphasize the importance of attention; for some it is the expanded and focused attention on all aspects of the present-now experience, as in mindfulness; for others it is the restricted and focused attention on the internal experience and a particular anchor, such as a mantra or image.

Recording the meditative process

Like GSR biofeedback, neurofeedback allows a meditator to visualize and modify an internal physiological process. It is not unheard of that people who believe they are meditating are actually quietly thinking about their plan of action for the day or reminiscing about a night spent with a friend. At the end of the session, they may feel more anxious or emotional. Examining an EEG recording of the session would reveal a significant amount of beta wave activity, which would indicate that thoughts were flitting through the frontal cortex. On the other hand, meditators sitting in presumed deep meditation may actually be in the early stages of sleep, otherwise known as 'holy dozing', and feel groggy and disorientated afterwards. An EEG recording would reveal a prominence of delta wave activity.

Although it is undoubtedly difficult to interpret EEG recordings accurately, it is possible over time to correlate various subjective internal states and specific changes in an EEG pattern. In this way, meditators can gain a better understanding of their state of mind while they meditate and discover ways in which their technique can be improved. Most experienced meditators would be aware, for example, if their mind were awash with thoughts, but for novice meditators, neurofeedback can facilitate the process of fine-tuning their technique and becoming more aware of their internal state.

Neurofeedback can also train a meditator to make a conscious switch between different internal states. We saw earlier how this can be used to optimize the performance of the brain, and it can also be used to facilitate meditation. A meditator who undergoes neurofeedback training will gain over a period of time the ability to switch quickly and efficiently, for example, into a state in which alpha waves are predominant, that is, a state of alert awareness.

Like the GSR meter, neurofeedback can also be used to assess how effective different contemplative techniques, such as focusing on the breath or repeating a mantra silently, are for any one individual. This minimizes the 'trial and error' period of exploring different techniques in an attempt to find the most suitable. Obviously, the nature of neurofeedback, and indeed biofeedback, restricts their use in contemplative practices involving movement, such as tai chi or yoga, but all contemplative practices have a meditative element that can be fine-tuned with the help of these technologies.

It may seem, then, that neurofeedback could provide the ideal aid to meditation. Is it really so easy?

The way forward?

Maxwell Cade's correlation of levels of consciousness, and different psychological stages of meditative practice, with different brain wave patterns prompted research into the possible use of neurofeedback to facilitate meditation. However, these attempts have been hampered by the inherent variability of the EEG recordings. Not only can EEG recordings differ both between individuals and within a single individual at different times, but different types of meditation also produce highly variable changes in the EEG brain wave pattern, both in the short term and the long term. Meditation cannot be viewed as a distinct state, as Cade suggested; it can only be viewed as a dynamic unfolding of different brain processes, and therefore a chain of brain wave changes. Even if it were possible to capture the full, dynamic changes in brain wave activity during a meditation session of a single Zen monk, for example, it would still not guarantee a template that could be used successfully to trigger the same experiences in another individual. Everyone is unique; no one template fits all.

It is also apparent that mystical experiences, whether natural or drug-induced, need to occur within a stable framework to allow for the knowledge that they convey to be integrated into everyday, left-brain function. The insights gained from a mystical experience require a certain level of psychological maturity. Just as entheogenic experimentation can leave some individuals disorientated and in need of psychological and/or spiritual guidance, technology-induced altered states of consciousness can also leave some people vulnerable, confused and unstable if they do not receive the appropriate support.

For the 'we want it, and we want it now' generation, the fact that neurofeedback has failed to offer the easy route to enlightenment has dampened interest in the application of this technology to mystical endeavours. However, continued support for neurofeedback, particularly in the USA, reveals that many still believe that this technology can enhance, or even optimize, brain function, as well as facilitate the practice of meditation. For some, this process of optimization is viewed as the next step in our evolution as a species.

It is perhaps fairest to say that the most viable strategy for anyone intent on gaining self-knowledge and a greater awareness and control over their behaviour may be the combination of contemplative techniques, developed and fine-tuned over many centuries, in conjunction with what recent technological advances have to offer. American psychologist Charles Tart had it about right when he wrote: 'I suspect that the more sophisticated use of biofeedback techniques will eventually aid basic meditative practices, but I doubt that it will substitute for it.'[7]

Meditation and health

WE HAVE EXPLORED HOW AND WHY MEDITATION WORKS, AS WELL AS POSSIBLE WAYS OF FACILITATING MEDITATIVE PRACTICE. However, we also need a better understanding of its short- and long-term impact on our health and wellbeing, and the role meditation can play, both in complementing more traditional therapies in patient care and as a means of maintaining good health.

Media interest in meditation has grown tremendously in recent years, but media coverage tends to present headline information and, in some instances, misrepresent or incorrectly interpret study findings. Therefore this chapter looks at some of the important studies conducted over the last five to ten years into the impact of meditation on health.

Health benefits of meditation

The exact physiological effects of meditation differ tremendously from one person to the next, and also depend on the experience of the meditator and the specific meditative discipline. However, generally speaking, meditation can elicit profound benefits in terms of physical, psychological, emotional and spiritual wellbeing.

In an urban study in 2004 of mindfulness-based stress reduction (MBSR), an eight-week programme showed significant improvement in participants' emotional and social functioning, as well as their general health and vitality. Although no impact on physical functioning was detected, the participants found themselves more capable of working and performing their activities of daily living. These improvements were reflected in an overall improvement in their health-related quality of life.[1] Generally speaking, the size of these benefits tend to increase with more frequent meditative practice, and are most pronounced in experienced meditators. These benefits provide clues about the possible therapeutic applications of meditation in healthcare.

Stress levels

Meditation has been shown to decrease cortisol levels in healthy volunteers and in patients with cancer. In Chapter 1, we saw that cortisol is a hormone released during periods of stress, so a decrease in cortisol levels reflects a decrease in a person's stress levels. Long-term practitioners of transcendental meditation (TM) have been shown to have significantly lower levels of cortisol than control subjects.[2] Furthermore, TM has also been reported to reduce levels of the hormones adrenaline (or epinephrine) and noradrenaline (or norepinephrine), two hormones associated with the sympathetic nervous system and our 'fight or flight' response – the body's response to physical and psychological stress.[3]

MBSR can improve physical symptoms and psychological distress, as well as improve general wellbeing. An analysis of 20 published and unpublished studies of MBSR included studies in a wide range of different patients (pain, cancer, heart disease, depression and anxiety). Overall, these studies indicate that MBSR is an effective method of stress reduction, associated with clear benefits in terms of overall health and in enabling ill people to cope with their condition.[4] MBSR has also been shown to have positive therapeutic effects in medical students, who commonly report high levels of stress and low levels of psychological wellbeing.[5,6] The students were less anxious and reported less psychological distress, including depression, as well as increased empathy. Results from another recent study showed that the number of visits made by urban patients to their healthcare provider was substantially reduced following the completion of a MBSR training programme.[7]

The positive impact of meditative techniques such as MBSR and TM is also evident in their beneficial effect on stress-related medical conditions. For example, the combination of conventional treatment and MBSR produced complete clearance of skin lesions in patients with the stress-related skin disorder psoriasis in a much shorter period of time than conventional therapy alone.[8] Similar benefits have been found in people with asthma. Emotional stress can both trigger and exacerbate asthma, and stress-reduction techniques can be a useful adjunct to conventional therapy. In an early study, about two-thirds of asthmatic patients reported improvements in their asthma with TM training.[9] Benefits of meditation have also been reported in patients with the stress-related irritable bowel syndrome (IBS): a six-week relaxation response meditation programme produced significant improvements in flatulence, belching, bloating and diarrhoea.[10]

The effects of meditation on levels of stress can also be detected as changes in the GSR. This, as we saw in Chapter 5, is a measure of skin resistance related to the level of arousal in the brain. Meditation has been shown to trigger an increase in the GSR (i.e. relaxation) that stabilizes after five to ten minutes. Similar increases were also detected while listening to music, but only when the music matched the individual's personal musical preferences.[11]

Meditation also has an effect on stress-inducing behaviour. People receiving training in meditation, followed by three-and-a-half weeks of regular practice, reported having an altered perception of time that not only reduced impatience, but was also found to reduce feelings of hostility triggered by periods of enforced waiting.[12]

The immune system

The stress hormone cortisol is also known to suppress the immune system. People with chronically high cortisol levels are therefore more susceptible to infection. Mindfulness meditation has been shown to boost the immune response to a vaccination,[13] and meditation has also been shown to reduce the body's immune response to stress, such as that evoked by prolonged, strenuous activity.[14] A cognitive behavioural stress management (CBSM) programme that included a number of different relaxation techniques, including meditation, has been shown to be of benefit in patients with HIV: patients exhibited increased immune activity after ten weeks on the programme. This improvement is thought to be the result of both a reduction in stress levels and reduced levels of depression.[15]

In several studies, chi kung has been reported to enhance the immune system's functioning. Chi kung exerts an important influence on the immune system in two main ways. First, the movement style of chi kung is thought to stimulate the immune system – short periods of moderate exercise have been shown to improve immune function, whereas longer periods have been reported to suppress it. Secondly, the psychological component of chi kung is thought to modulate levels of stress, and thus cortisol. Since cortisol inhibits the activity of the immune system, this drop in cortisol has the effect of boosting the immune system.

Melatonin and cancer

Melatonin, which is produced by the pineal gland at night, is responsible for generating the biological rhythms that dominate our existence, from the diurnal rhythm that dictates our sleeping pattern to the rhythm that defines our reproductive cycles. It has also been suggested to have a number of other roles, including a role in the development and treatment of several types of cancer, particularly those with hormonal involvement, such as breast and prostate cancer.

Melatonin has been shown to enhance the activity of the immune system, as well as suppressing the growth of breast cancer tumour cells grown in cultures in the laboratory.[16] Although there is still a need for more evidence, a preliminary study revealed that patients with conventionally untreatable, advanced tumours showed considerable improvement after treatment with melatonin; disease control was noted in about 60 per cent of patients with an otherwise poor prognosis.[17]

People who meditate regularly have been shown to have

higher levels of melatonin than those who don't. It has been demonstrated, for example, that experienced meditators, trained in either TM-Sidhi or another internationally recognized yoga discipline, have much higher levels of melatonin following a night-time meditation session than those seen on a control night on which they didn't meditate.[18] The exact way in which meditation induces this increase in melatonin levels is not known, although it has been suggested that meditation slows blood flow through the kidneys, therefore prolonging the period that melatonin is present in the blood before being metabolized and excreted by the kidneys. In another study, regular meditative practice was shown to significantly increase melatonin levels in patients with breast cancer.[19] There are few therapeutic strategies that effectively increase melatonin levels, so the observation that meditation can increase melatonin suggests yet another possible therapeutic benefit of meditation and a possible role for it as a useful adjunct to conventional anti-cancer therapies.

In a study of the effect of MBSR in patients with breast and prostate cancer, eight weeks of MBSR practice resulted in improvements in the symptoms of stress experienced by the patients and a general improvement in quality of life.[20] Furthermore, although the changes in their immune function were not significant, there were indications of a slight positive shift.

The strongest evidence for the use of meditation in patients with cancer stems from a study that showed that MBSR led to a 65 per cent improvement in mood and a 35 per cent reduction in stress symptoms in a mixed population of cancer patients.[21]

Cardiovascular health

Much of the impact of meditation on the health of the heart and blood vessels is likely to be a consequence of reducing stress levels. Both heart rate and blood pressure rise during periods of stress, so it is perhaps not surprising that meditation can lead to notable reductions in the risk of cardiovascular disease (CVD).

Heart rate changes

Studies have shown that a lower heart rate is not only associated with a decreased risk of death from CVD, but also reduces the risk of death from all diseases.

In Chapter 3 we saw that meditation is superimposed over a general relaxation response, which slows the heart rate and breathing rate. However, the lowering of the heart rate during meditation is not solely attributable to this relaxation response. A study in 2004 found that the heart rate of experienced meditators is significantly lower after a session of meditation than that of study participants who merely rested during the same period.[22] Other studies reveal that certain types of meditation can also be associated with the heart rate fluctuating both up and down in a slow rhythmic fashion. During these oscillations, the heart rate can change by as much as 20 beats a minute and the frequency of these oscillations matches the breathing rate – as a person inhales, the heart rate increases, and the heart rate then decreases on exhalation. When practising 'breath of fire' meditation, which entails breathing rapidly through the nose, significant increases in mean heart rate compared with baseline were detected.[23]

In the long-term, meditation can lead to reductions in the baseline heart rate and an increased control over the heart rate. In a study of novice yoga practitioners, 30 days of yoga training not only significantly lowered the baseline heart rate of participants, but gave them an enhanced ability to lower their heart rate voluntarily.[24]

Hypertension (high blood pressure)

There is evidence to suggest that meditators have lower blood pressure, both before, during and after a session of meditation.

Regular, long-term meditative practice can reduce the blood pressure of borderline hypertensive patients, and also protect people who risk developing hypertension. Meditation should not be considered to be an appropriate substitute for pharmacological management of patients with high blood pressure, but it is a promising add-on therapy.

In a recent study, patients with mild to moderate hypertension were randomly assigned to two treatment groups. The first group received instruction in contemplative meditation with breathing technique (CMBT) and followed an eight-week practice schedule. The second group did not receive CMBT, and acted as a control. Those in the CMBT group were found to have lower blood pressure compared with the control patients, both at rest and during exercise.[25]

The feasibility of MBSR as a treatment for high blood pressure has only been explored in a limited number of studies, despite an overwhelming body of anecdotal evidence of its positive effect in lowering blood pressure. One study of healthy, adolescent students showed MBSR leads to significant reductions in blood pressure, both at rest and during the day at school,[26] and other studies are ongoing.

Cardiovascular disease

A number of rigorous studies in recent years have indicated that long-term TM effectively reduces CVD and the risk of developing it. This is thought to be a reflection of meditation's influence on a number of different risk factors for CVD: not only has TM been reported to decrease blood pressure, reduce the intake of tobacco and alcohol and lower cholesterol levels, but, it has also been suggested, decreases psychosocial stress. Long-term TM has been associated with a reduced incidence of atherosclerosis (the build-up of fatty deposits on the inside walls of the arteries) and myocardial ischaemia (an imbalance between the heart muscle's oxygen supply and demand), and a reduction in CVD-related deaths. Again, these benefits are largely probably the result of meditation's beneficial effect on stress levels.

Cognitive changes

Meditation is a method of training attention, and so its practice is often associated with improvements in cognitive skills. Unsurprisingly, regular meditation improves the ability to maintain single-focused attention, even in the face of internal and external distractions. Meditation can also improve problem-solving and decision-making skills. These improvements stem from a detachment during meditation from conditioned ways of perceiving and interpreting experiences that allow a person, later, to explore the problem or decision from a different angle.

A review in 2003 of studies investigating the cumulative effects of TM on cognitive function identified 107 articles that investigated this issue.[27] Of these, only ten were found to meet fairly strict criteria with respect to their methods. Four of these reported a positive effect of TM on cognitive function. All the participants recruited for these studies were predisposed towards TM, and this evidence has been interpreted by some to mean that the observed benefits of TM were a result of their inherent expectations. However, it can also be interpreted to demonstrate the importance of intent where meditation is concerned. Someone intent on learning meditation, and intending to practise it regularly, is more likely to adhere to regular practice and therefore more likely to derive the benefits.

A number of long-term studies included in the review have also evaluated the effect of twice-daily TM practice on general measures of intelligence. In the most recent of these, participants who practised TM showed greater improvements in general intelligence than were seen in the control group.

Psychological impact

Meditation has also been associated with a number of subjective psychological effects. Regular meditators report some or all of the following:

- a boost in energy levels: a decreased need for sleep and daytime naps, an increase in productivity and creativity and in physical stamina
- increased self-acceptance and a gradual release from the tendency to attribute self-blame. This often also translates into an increased acceptance of other people and their eccentricities, which can improve interpersonal relationships
- a greater ability to express emotions, both positive (pleasure and love) and negative (anger and sadness). During meditative practice, suppressed memories and the emotions associated with them are often released, and they find themselves in tears or laughing uncontrollably
- fewer bouts of irritability and impatience, or emotional or behavioural outbursts
- an improved and expanded sense of identity
- a greater understanding of which situations, individuals and behaviour are constructive, and which are destructive. This may also enable them to express their opinions with greater confidence and be able to make decisions more quickly.

Meditation has accordingly been shown to have a positive effect on the outcomes of psychotherapy. A study in 1985 reported, for example, that patients undergoing long-term, individually tailored psychotherapy responded well to a ten-week meditation programme. Both the therapists and the patients noted that meditation brought about significant and substantial improvements in the patient's psychological wellbeing.[28] It is important to recognize, however, that the response to meditation can depend on the particular aims of the therapy sessions. If the aim is to evoke cognitive or behavioural changes, meditation can be a useful adjunct to more conventional psychotherapy. If, however, the aim is to reinforce the ego boundary, as in disorders of the self (such as schizophrenia or multiple personality disorder), or to release powerful buried emotions or tackle complex relational issues, conventional psychotherapeutic approaches are thought to be more effective, and meditation may even be counterproductive. In general, patients with higher levels of psychological distress respond less well to regular meditative practice.

The investigation of the psychological effects of meditation is particularly difficult, since there is evidence that meditators and non-meditators may differ in ways that cannot be attributed purely to meditation. Some early studies have shown, for example, that people who decide to learn to meditate tend to be slightly more neurotic and anxious than average. This does not negate the benefit derived from meditation – not only are long-term meditators less anxious than novice meditators and non-meditators, but novice meditators show significant decreases in measures of anxiety following meditation.[29, 30] In one long-term study, patients with anxiety disorders showed significant improvements in both subjective and objective measures of anxiety levels following three years of meditative practice.[31] Similarly, meditative practice has been reported to reduce levels of neuroticism related to the frequency of meditative practice: only regular meditators exhibited a significant decrease in neuroticism.[32, 33] However, all these studies are more than 20 years old. There is still a notable lack of robustly designed studies into the effect of meditation on patients with anxiety. A meta-analysis in 2007 considered only two published studies of an appropriate quality to include in the analysis and so, unsurprisingly, its analysis did not support the use of meditation in patients with anxiety.[34]

There is, however, emerging evidence of a benefit of mindfulness-based cognitive therapy (MBCT) to patients with depression and anxiety. People who are depressed often find themselves ruminating about their personal shortcomings and problems. With all of these negative thoughts running through their head, they often fall even deeper into depression. MBCT teaches them to acknowledge these negative thoughts and then let them go. When a negative thought arises, they learn not to allow themselves to become absorbed in the thought and the negative emotion associated with it, neither berating themselves for having the thought or attempting consciously to block it. Instead they learn to accept the thought for what it is, a mere thought, without viewing it in terms of its content or emotional connotations.

This attitudinal shift has been linked with marked improvements in depression. Patients with anxiety and depression undergoing a three-month programme in MBCT reported significant improvements in their emotional state, and more than half continued with the mindfulness-based technique after the programme.[35] In another, more robustly designed study, patients recovering from an episode of depression were randomly assigned to receive their standard treatment with and without MBCT. MBCT was found to significantly reduce the risk of depressive relapse in patients who had had three or more previous episodes of depression (which was more than three-quarters of the participants).[36]

Combating addictive behaviour

We have already seen that TM has been suggested to reduce alcohol and nicotine consumption; regular, long-term practice TM has also been noted to have a beneficial effect in the struggle to reduce the use of illegal substances, tranquillizers, prescribed medications and even caffeine.[37] The decrease in drug use was seen to be greatest in those who participated most fully in the meditation programme and for the greatest length of time.[38]

An eight-week MBSR programme showed promise as a new

method of quitting smoking, with more than half the participants still abstinent six weeks after completing the programme.[39] This group was largely made up of people who meditated for about 45 minutes a day, compared with the 20 minutes a day reported by non-abstainers. Successful abstainers also experienced less stress and emotional distress from giving up smoking. Interestingly, the people who were most interested in meditation at the start of the programme were the ones most likely to remain abstinent for the duration of the study.

These reported reductions suggest a decreased reliance on external means of altering the physical and mental state. A growing body of evidence suggests that smoking and drug taking are associated with depression. By relieving some of a person's stress and emotional distress, meditation can reduce the urge to smoke or take drugs. Meditation is also likely to lessen the attention paid to intrusive thoughts that, although often transient, can drive an urge for an addictive substance.

The impact of meditation on addictive behaviour varies tremendously from one person to the next, but even limited success puts it on par with other aids currently available.

Coping strategies

Chronic pain is an ever-increasing clinical, social and economic problem, and in recent years there has been increased emphasis on possible supportive techniques that can be taught to sufferers.

Mindfulness meditation, which develops a state of detached self-awareness, is a promising support for patients. Jon Kabat-Zinn, who developed MBSR, reports that 65 per cent of patients with chronic pain who had failed to respond to traditional medical care showed marked improvements in pain after ten weeks of MBSR. The majority also showed improvements in mood disturbances and the number of troublesome symptoms.[40] Exploring these benefits in greater detail found that mindfulness meditation improved patients' body image and self-esteem and their ability to carry out everyday activities, lessened psychological problems such as anxiety and depression, and reduced patients' consumption of pain-relieving medications.[41]

Bone marrow transplant recipients are frequently distressed, depressed and anxious, and in need of psychological and emotional support during treatment. Current support falls short of patient needs, and meditation represents a promising way forward. One eight-week MBSR programme produced improvements in mood and reduced levels of depression and anxiety.[42] These improvements failed to reach statistical significance, which is perhaps unsurprising considering the small number of participants and the short-term nature of the MBSR programme, but the results warrant further investigation in larger, more robust trials.

Possible unpleasant side effects of meditation

Meditation may not be appropriate in patients with certain mental health disorders. Patients with disorders of the self, such as those with schizophrenia, interpret their mystical experiences differently from so-called 'normal' people. Mystical experiences often involve a widening of the sense of self and, in the absence of a fully functional and unfragmented ego, this awareness can exacerbate the problem.

Meditation is also questionable in epileptic patients. People who suffer from disorders of cortical excitability, such as epilepsy, experience mystical experiences more frequently than average, and in an uncontrolled fashion. Some researchers maintain that meditation can be beneficial in epileptic patients, as it teaches greater control over the activity in their brain, whereas others suggest that meditation can exacerbate the symptoms and is therefore potentially very harmful. Care is obviously warranted until this fierce debate is finally resolved.

Many people wrongly assume that meditation is solely

associated with positive experiences. However, as with psychotherapy, the cognitive and behavioural changes triggered by regular meditative practice can produce a wide range of emotions, from fear and anxiety stemming from a gradual loss of the ego's defences, to sadness or anger from the emergence of repressed and often traumatic memories and images. It is not always appropriate to express our emotional reaction to a particular experience at the time, but when these emotions are particularly strong they rarely dissipate of their own accord. Instead, they are buried deep in our unconscious. The experience during meditation is very personal in nature. In the early stages of the process meditators access their personal unconscious, and become aware of images and memories not normally accessible to their conscious mind. The surfacing of these images and memories is often accompanied by the surfacing of the emotions associated with them, and it is not unusual for practitioners to find that, during meditation, they find themselves weeping uncontrollably or bristling with anger.

As with conventional psychotherapy, these emotional releases are a necessary part of the process and should not be a deterrent to meditative practice. However, some people learning to meditate may derive benefit from seeking guidance from a psychotherapist or counsellor, to deal appropriately with these 'birthing pains'. Inevitably, access to deeper and more expansive levels of consciousness will reveal suppressed pains or 'skeletons in the closet', but through these negative experiences there is the opportunity for growth. If repressed emotions are acknowledged and accepted, they are released, with a resulting release of tension. If, on the other hand, these emotions are resuppressed, a build-up of tension can manifest in depression, anxiety and ill-health.

Some apparently positive experiences of meditative and mystical experiences can also be a hindrance to true personal development. The potential pitfalls can be illustrated by any one of countless examples of so-called 'gurus', such as Sai Baba or Da Free John, whose experiences during meditation inflated their self-esteem, conferring delusions of grandeur and encouraging them to engage in promiscuous sexual behaviour.

The worldview stored in our left brain is very ingrained; it represents the sum of all our experiences. We cling to this as it appears to provide certainty and security in an inherently unpredictable universe. The unknown is always scary, and experiences that appear to undermine our ego are seen as direct assaults on our chances of survival. The left, storytelling side of our brain will therefore try to devise an explanation for our experiences that fits within our limited worldview. In mystical traditions, wisdom derived from meditation is viewed as a gift from God; a source of knowledge that embraces but transcends the personal self. In the absence of some kind of spiritual mental map, a person can find the knowledge gained through meditative experiences difficult to integrate into everyday life. This mental map, although abstract, allows knowledge arising from right-brained activity to be integrated into the left-brained worldview. Without this selfless approach, the ego can hijack this wisdom, claiming ownership over it, and thus often lead to an inflated ego and its inherent risks.

A study carried out in 1992 explored the possible adverse effects of meditation in long-term meditators.[43] Participants were asked to report adverse effects that they had experienced in the past, as well as those experienced one and six months after a meditation retreat. About two-thirds of them reported that their meditative practice had been associated with at least one adverse effect. These included relaxation-induced anxiety and panic, decreased motivation, confusion and disorientation, depression, feeling 'spaced out' and paradoxical increases in tension. However, they also reported that, both retrospectively and six months after the retreat, the positive effects of meditation outweighed the negative effects.

These adverse events do not by any means occur in all meditators, or within a single person with any frequency.

Subjective reports of confusion, depression or anxiety related to meditation also fail to take into account the countless other factors that can influence an individual psychological state during the course of a day, or even an hour. Undoubtedly, non-meditators also experience moments of anxiety, depression, disorientation and tension. It is the frequency with which these occur that is the most important, not the fact that they occur. This has yet to be investigated in a controlled study.

Studying the impact of meditation on health

The studies and programmes that have contributed to the information above reflect the dramatic increase in the study of meditation in recent years. Just five or so years ago, a search of an electronic database of all the research papers published in reputable journals would have revealed a handful of credible studies. Now, a similar search produces more than 1,300 research papers stemming from budding research centres across the Western world. We are still, however, at the beginning of our learning curve. In other areas of medical research, in which our knowledge of a disease and its treatment has a firm foundation in decades of research, an electronic search produces hundreds of thousands of research articles. Our current understanding of the physical and psychological benefits of meditation holds much promise but is merely the tip of the iceberg. However, gleaning meaningful information from this growing body of research is fraught with difficulties and results are prone to misinterpretation.

Pitfalls in studies of meditation

A first problem is difficulty in assessing accurately how strictly a study participant adheres to the programme. There is no guarantee that all study participants practise with the same frequency, or indeed for the entire duration of the study. In one study, for example, 58 per cent of study participants reported that they practised meditation; however, only 37 per cent were found to be practising regularly.[44] Non-compliance with the programme may prevent the study from detecting any significant health benefits of meditative practice, and it may also explain why different studies often report conflicting findings. Whereas the study of a new drug, for example, involves the controlled administration of a specified dose of the drug, it is impossible to provide a 'fixed dose' of meditation.

It is also difficult to apply the same rigorous methods routinely used, for example, in investigating pharmaceuticals to the study of meditation. It is impossible to impose homogeneity across the participants as each recruit will inherently differ in terms of psychological, emotional as well as spiritual outlook. Some participants may be more inclined towards meditation, whereas others may be in a psychological state in which meditation is likely to be ineffective or even detrimental to them. There are some noticeable differences between short-term and long-term practitioners of contemplative disciplines. It is therefore crucial that studies recruit volunteers with a wide range of prior experience for the findings to be applicable to the general population, rather than just experienced meditators.

For some people, these inherent limitations question the validity of the findings of studies of meditation, in the same way as the validity of findings of psychological research has traditionally been challenged. However, just as the use of such rigorous experimental methods in the study of psychological treatments is questionable, so they are in the study of meditation. Meditation is not meant to replace conventional healthcare. New pharmaceuticals must be tested extensively and rigorously before being released, as these products claim to offer benefits for a particular condition that often requires urgent treatment. Inadequate treatment can have serious consequences for the patient. Meditation, on the other hand, is associated with benefits in terms of overall health and can therefore be used as a preventative or add-on therapy rather than an acute treatment. Its study should not therefore be subjected to the same rigorous criteria applied to investigations into acute treatments.

Another consideration is that long-term studies are difficult and costly to run. Anecdotal evidence suggests that the health benefits of meditation emerge after continuous and intensive practice over a period of several months, if not years. In order to best detect the benefits of meditation, study participants must be subjected to equally intensive training and practice. The current lack of long-term information about the impact of meditation will be soon addressed by the Shamatha Project. This project, run by Alan Wallace at the Santa Barbara Institute for Consciousness Studies, aims to be a full exploration of the neural, cognitive and socio-emotional effects of intensive training in a Buddhist technique called shamatha meditation. Volunteers with varying degrees of prior experience will be randomly assigned to participate in one of two three-month retreats. It is hoped that the findings of this study will shed more light on the benefits of long-term meditation.

Integrating meditation into healthcare

Evidence suggests that meditation can have a measurable effect on certain aspects of health. Meditation can be a useful adjunct to treatment of stress-related diseases, or diseases exacerbated by stress, and it has also been shown to have promising effects in patients with cancer and those dealing with the burden of chronic disease. This presents an argument for meditation's integration into patient care. Our healthcare systems are struggling to deal with the needs of an increasing unhealthy population; solutions are needed, and needed quickly. Meditation promises to offer that solution, improving our general health and reducing our current reliance on the healthcare system to repair the damage inflicted by our fast-paced, stress-filled lives.

The trouble with patients

The changing landscape of healthcare hinges on the pivotal role of patients themselves in the prevention and management of disease. Nowadays, we have a much better understanding of the risk factors for certain diseases and can, in many cases, prevent them from developing in the first place. However, we tend not to avoid these risk factors consistently. Countless people still smoke, despite increasingly horrific pictures and written warnings on cigarette packaging. Countless people still eat junk food and choose to flop on the sofa rather than take a walk. People with type II diabetes are usually advised to avoid excessive sugar, alcohol and smoking; however, only a worrying small proportion will actually manage to stick to this advice consistently. It is part of human nature to want to do what we are not meant to do.

Compliance with medication is also always poor. Some patients forget to take their medication at the prescribed times, others avoid it because of troublesome side effects or the belief that it is not working. No matter how sophisticated and well-informed a treatment plan, it will fail unless the patient accepts responsibility for his or her own role in implementing it.

The benefits of self-help

The increasing role played by patients is slowly being acknowledged, both by the medical profession and by patients themselves. This is reflected in a more patient-focused approach to health and ill-health, and also in the emergence of patient self-management programmes that aim to both encourage patients to become informed about their condition and to promote their active involvement in the management of their own health. The aim of these self-management programmes is to create a pool of so-called 'expert patients' who feel empowered to take control of their lives and their disease and use the knowledge they gain to best manage their condition. Expert patients are more likely to have positive treatment outcomes, reaping the benefits of a close and communicative relationship with their healthcare provider, while remaining realistic about the impact of their disease on themselves and their family.

Integrated care in action
The following are just three examples of how meditation is forming an integral part of healthcare.

MBSR programmes
There are hundreds of different MBSR programmes currently running in North America and Europe, many of them affiliated with prominent medical institutions. The University of Massachusetts Medical Center (UMMC) for Mindfulness in Medicine, Healthcare and Society runs the oldest and largest MBSR programme in the USA. More than 13,000 individuals have received training since the centre was founded in 1979, and it maintains a database of trained MBSR practitioners offering training programmes across the globe.

The MD Anderson Wellness Programme
The MD Anderson Cancer Center in Houston Texas is one of the premier oncology centres in the world. It offers more than 75 complementary therapy programmes, including instruction in Buddhist meditation, centering prayer, yoga, tai chi and chi kung. These complementary therapies are provided in addition to standard care, to manage cancer-related symptoms, relieve stress and improve overall quality of life for patients. The Center has also developed a website providing educational resources about complementary therapies. This valuable repository of information about complementary therapies was designed to educate medical professionals and patients alike about the potential benefits and drawbacks of these therapies, promoting their integration into cancer healthcare.

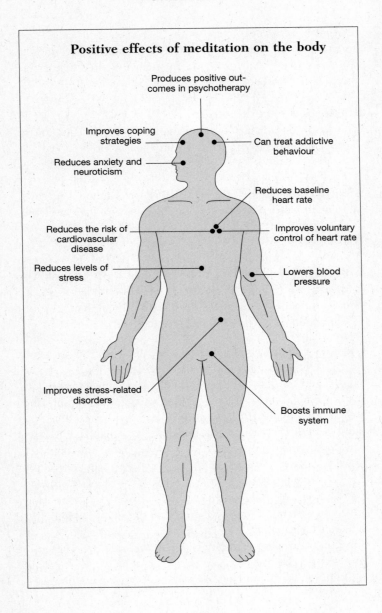

Positive effects of meditation on the body

Produces positive outcomes in psychotherapy

Improves coping strategies

Can treat addictive behaviour

Reduces anxiety and neuroticism

Reduces baseline heart rate

Improves voluntary control of heart rate

Reduces the risk of cardiovascular disease

Reduces levels of stress

Lowers blood pressure

Improves stress-related disorders

Boosts immune system

The Oxford Cognitive Therapy Centre

In the UK, the Oxford Cognitive Therapy Centre offers a practical course for people with depression that combines MBSR techniques with cognitive therapy. This course, run by Mark Williams, teaches participants to become more aware of the early signs of an impending depressive relapse and, in some cases, to avoid relapse. In an ongoing pilot study run at the centre, Mark Williams and his colleagues are evaluating the use of MBCT in people who have attempted suicide, but recovered. This study aims to investigate whether MBCT can reduce the risk of these patients committing further self-harm. The centre also offers training for healthcare professionals interested in cognitive therapy in general, and MBCT in particular.

Summary

Studies into meditation are fraught with difficulties. Despite this, in recent years, there has been an increased focus on research into the health benefits of meditation and, as a result, evidence for a number of clear health benefits has emerged. Meditation appears to reduce stress levels, boost the immune system, improve outcomes and the quality of life for people with cancer, and reduce the risk of cardiovascular disease. It also has important effects on psychological wellbeing, enabling people to cope better with ill-health and often eliciting substantial psychological, emotional and spiritual growth.

Meditation should be viewed as an essential component of any self-management programme. A relatively small amount of training – only eight weeks in the case of MBSR – equips people with the knowledge needed to start integrating meditative practice into their everyday lives. Combined with beneficial lifestyle changes, and standard pharmacological or surgical treatment, meditation can offer patients improved physical, psychological, emotional and spiritual wellbeing. This multi-disciplinary approach to patient management promises to improve both treatment outcomes and patient satisfaction, and will undoubtedly, in the long term, relieve some of the burden currently placed on our struggling healthcare systems.

Meditation in our everyday lives

IN ORDER TO GET A PROPER FLAVOUR OF MEDITATION, IT IS CRUCIAL TO SPEND TIME WITH A TEACHER, TRAINING UNDER SUPERVISED CONDITIONS. It is impossible to capture all the elements of meditation in a few pages of text. However, I hope that this last section of the book, dedicated to a short practical overview of two forms of meditation, will provide a helpful insight.

Mindfulness-based stress reduction (MBSR)

MBSR trains you to develop enhanced awareness of the right-brained, moment-to-moment experience of emergent mental processes. Unlike traditional forms of cognitive therapy, MBSR does not aim to change your way of thinking, merely change your attitude towards these thoughts when they arise. It involves non-judgemental awareness of both internal manifestations (thoughts, memories and emotions) and external events (sounds, sights or other sensations). These are all viewed in a neutral manner, with no attempt to extract meaning from them or to analyse them.

MBSR also involves present-moment focus. In everyday life, there are usually many different stimuli requiring our attention at any one moment in time. In my office, part of my attention might be focused on an article I am writing, part focused on answering the query of a work colleague about an unrelated project, part on the alert that has appeared on my computer announcing an email from my boss. It is rare that I can give something my undivided attention. Even our thoughts provide distractions. When sitting in the park, my mind has a tendency to wander, remembering events from the past and hypothesizing about the future. MBSR trains you to bring your awareness to the present moment; to the task or experience at hand. MBSR also teaches you to become more aware of automatic behaviour – often we act without thinking, a behavioural 'knee-jerk' reaction. MBSR inserts a pause into this behavioural reflex, allowing your reactions and behaviour to become more considered and deliberate.

The MBSR programme devised by Jon Kabat-Zinn acknowledges three meditative traditions. First, it involves seated meditation and a series of meditative breathing exercises. Secondly, it involves a body scan meditation, systematically drawing attention to each specific area of the body. Thirdly, it involves gentle and strengthening yoga postures, combined with a present-moment awareness of the breath and the sensations associated with adopting the different postures. MBSR has a number of defining features.[1]

Non-judgemental attitude

We all have the tendency to rate our experiences and thoughts into different categories, each of which is associated with a pre-programmed response. A so-called 'bad' experience therefore leaves us feeling depressed and anxious, whereas a so-called 'good' experience leaves us feeling elated and optimistic. These pre-programmed responses take us on an emotional roller-coaster. MBSR teaches you to view all experiences and thoughts impartially and non-judgementally and, in doing so, provides a way to get off this emotional roller-coaster. Rather than interpreting experiences, or thoughts, as being 'good' or 'bad', or 'exciting' or 'worrying', you merely observe them without judgement.

Patience

In our fast-paced society, we often expect results quickly and become impatient if we have to wait for anything for more than a short space of time. However, being impatient does not accelerate events, it merely creates anxiety and frustration. Furthermore, when the moment finally comes, we may find that it was not what we were expecting and it doesn't bring the benefits or rewards we assumed it would. This of course leads to disappointment. Similarly, novice meditators can often become frustrated at their rate of personal development, or lack of it. They become impatient with themselves and, in doing so, actually hinder their practice. MBSR teaches acceptance of the natural pace of events and, in doing so, cultivates greater patience.

Beginner's mind

The term 'beginner's mind' refers to a receptive and open attitude, a willingness to see things are they are rather than how we think they are. By the time we reach adulthood, many of us have developed fairly firm ideas about life, the universe and everything. This can leave us closed off to the new experiences, thoughts and behaviour that are so essential to our personal development. MBSR encourages the cultivation of 'an open mind', experiencing everything as if it were occurring for the very first time with no preconceptions or expectations. In this open mind there is space for new ideas, new ways of thinking and behaving; a readiness to see things from a different, less restrictive perspective.

Trust

An integral part of MBSR is building up trust in your inner voice or intuition. This trust is dependent on the inner confidence that we are in fact seeing things as they are rather than how we think they are. In our modern-day world, we are often reluctant to openly rely on 'gut instinct' or a 'hunch' – we need hard evidence that proves our theories are correct and therefore that the outcome will be as we expect. However, our intuition can be a powerful tool, and it can often be a more valuable guide than external sources. MBSR trains you to become more aware of your inner voice and how to distinguish it from the expectant or fearful voice of the mind. You learn to trust this inner voice and use it to guide your actions in everyday life.

Non-striving

We are taught from an early age that, in order to be successful, we need to apply ourselves with purpose – doing nothing is considered lazy. As a result, most of the things we do have an ultimate goal: we go to the gym to lose weight, we work hard to earn a promotion. MBSR, like all other forms of meditation, is by nature not goal-oriented. It draws attention to the present moment, observes it in a non-judgemental way, and does this in an unforced manner. There are many proven health benefits of MBSR; achieving these benefits cannot be a goal, only a possible by-product of regular practice. Embarking on an MBSR programme with the specific intent of, for example, lowering your blood pressure, will cause a natural attachment to that goal. If your blood pressure remains the same, you are likely to berate yourself or question the method. Such goal-directed behaviour fails to promote the non-judgemental attitude and patience inherent to MBSR.

Acceptance

We spend too much of our lives refusing to accept situations as they are. There are countless examples of people living in denial – someone who stays in a destructive relationship; a person who continues to smoke despite advanced respiratory disease. Once we see things as they really are and not as we suppose them to be, the next step is acceptance. If, for example, you have a tendency to become anxious, the first step is to accept that you have anxious tendencies: this is who you are, in the present moment. MBSR teaches you to accept the way things are, and this acceptance stems from both a willingness to see things as they are and a trust that things will unfold as they should and at their own pace. This feature of MBSR is perhaps best illustrated with a quote from Max Ehrmann's 'Desiderata': 'whether or not it is clear to you, no doubt the universe is unfolding as it should.'

Letting go

When we assign emotional value to particular thoughts, we become attached to those thoughts and our minds are therefore reluctant to let them go. In the case of a pleasant thought, such as the memory of a good night out with a loved one, we replay this memory over and over again, revelling in reliving the pleasure we experienced. We also replay unpleasant thoughts, such as a worry about an impending deadline or interview, imagining over and over again possible scenarios and outcomes. In both these cases, our attachment to these thoughts draws us away from the present-moment experience. MBSR brings a greater awareness of the thoughts, behaviour and experiences to which we have become attached. Letting go involves acknowledging them but taking the decision to not become absorbed in them or dwell on them.

Mantra meditation

The practice of mantra meditation has three essential components. Firstly, consciously relaxing ('stilling the body'), then drawing attention to the breath ('focusing on the breath'), and finally consciously relaxing the mind by repeating a mantra ('stilling the mind'). These three components, and some useful exercises, are given below.[2]

Stilling the body

In our fast-paced society, we are so accustomed to being on the move, that doing nothing can seem unnatural and uncomfortable. Our bodies are usually poised to take action, and this manifests as a feeling of restlessness whenever we have moments of stillness. The first crucial step in meditation is to overcome this restlessness and consciously relax the body. The following exercise may be helpful:

Step 1 Put all your attention on your scalp, a major carrier of stress; put your total concentration on what you can feel in your scalp. You may want to move your eyebrows or your ears to become aware that you actually do have a scalp. Focus all your awareness there. How does it feel? What can you feel? Can you feel anything? Breathe in gently and breathe out and let go of all tension and stress there. Put your scalp at ease. Gently repeat this several times, always breathing in gently and breathing out long and relaxed and let go.

Step 2 Next, move your attention to your face, especially on the frowning points between the eyebrows, around the nose and mouth. We tend to frown, pinch our nose and purse our lips more than we know. Put all of your attention there. Become aware of what you do. Move the muscles in your face. It doesn't matter. No one is looking. Now really frown, scrunch up your face. Then widen your face, relaxing the muscles. Breathe in gently and let go of all the tension in your face. Just breathe out and let go. Breathing in gently and breathing out, long, relaxed and let go.

Step 3 Now put all your awareness on your mouth, tongue and jaws. Where is your tongue? Is it against the roof of your mouth? This is a sure sign of tension and readiness to talk. Let it drop. Let it fall gently to where it naturally wants to be. Are your jaws clenched? Let them go. Put your awareness on the point where your jaws are connected, just under your ears. Put all your awareness there and breathe in gently and breathe out and let go. Just let them go. The mouth is closed but totally relaxed.

Step 4 Put your awareness on your shoulders. How do they feel? Move them up and down to become aware of how you hold them. Breathe in gently and breathe them down. It is amazing, how far down they will go. Just keep your awareness there and breathe out and let go.

Step 5 If the head and shoulders are at ease, the rest of the body will follow. But check for any tension in your arms and legs, especially the calves of your legs. Our legs work all the time and often we hold them tensed and ready to move, even when it is not necessary. Just tense and then relax each of the muscles of your arms and legs in rolling succession, constantly with total awareness. Every time your attention goes, just gently bring it back to whatever part of the body you are focusing on, breathing in gently and breathing out and letting go.

Step 6 Check once more in your own time and at your own speed right over your body from your scalp to your feet. Put your body at ease. If there is any pain or discomfort, put your whole awareness there. Feel the pain. How does it feel: burning, nagging or sharp? Really become aware of it, acknowledge it, then breathe in gently to that spot and then breathe out and let go. Just let go.

Focusing on the breath

In many traditions, the breath is used as a focus to both relax the body and still the mind. When we are stressed, we tend to breathe faster and take shallower breaths: the rapid, shallow breathing associated with a panic attack can lead to dizziness and a loss of consciousness. The next stage of mantra meditation is to draw awareness to your breath, consciously taking regular, deep, unforced breaths. The following exercise may be useful:

Step 1 Focus on the breath. Really get to know your breath. Just focus on the sensations near the nostrils: feel it coming in cool and feel it going out warm. Feel the air passing over the hairs in your nostrils. Don't alter your breath, just watch it, coming in cool and going out warm. Just breathe.

Step 2 Take the awareness of your breath a little further. Feel the breath cool at the back of your throat as you breathe in and warm as you breathe out. Just feel it coming in your nostrils and passing at the back of your throat, cool coming in and warm going out. Just breathe.

Step 3 Now take the awareness of the breath a little further still. Feel it going into your heart region. Notice the top of your chest moving slightly, feel the air going in and going out, your chest going up and going down. You might even feel your heart beating. Focus your whole attention on the air going in and going out. Just breathe.

Step 4 Next, go even deeper. Feel the air filling the bottom of your lungs. Feel your lower ribcage expanding and contracting. Just enjoy the movement of your ribcage. Just breathe.

Step 5 Go even deeper. Your diaphragm is like an upturned cup. When you breathe in, the diaphragm goes down and the cup flattens and becomes an upturned saucer. And when you breathe out, it goes up and back to its cup shape. Breathe in, going down and flattening; breathing out and going up. Focus all your awareness on your diaphragm moving up and down. Just breathe.

Step 6 Now just breathe naturally and watch your breathing. How does it feel? Where can you feel the breath? Breathe naturally, in and out.

Stilling the mind

When the body is still and relaxed, we often rely on our mental landscape to feed our innate restlessness, getting lost in elaborate daydreams or trips down memory lane. The quieter the body, the more aware we are of the thoughts racing through our mind: plans for dinner, worries about the future, memories from the past. The next stage of meditation is therefore to still the mind. This is achieved in two main ways.

First comes a change in attitude towards your thoughts.

Usually, when a thought arises, such as a comment made by a friend over lunch the day before, it triggers a chain of further thoughts, such as the need to buy the friend a birthday present or a memory of the food you ate. Often, you can get lost in this train of thought, happy to follow the mind wherever it leads. During meditation, you acknowledge the thought, but then, instead of examining it properly, you merely file it away for future inspection. Your attention is not drawn away by the thought and, as successive thoughts are acknowledged and then 'let go', the mind quietens and fewer thoughts emerge.

The second way of stilling the mind is to use an anchor to draw attention away from any emerging thoughts. This could be merely the breath. As soon you become aware of the fact that you have let your mind wander, you can draw attention back to your breath and, in doing so, release the thought. In other cases, a tool such as a mantra is used to focus the mind. A mantra is a word or phrase of spiritual significance that helps you to leave your thoughts behind and enter the silence. By repeating the mantra, you are less tempted to become distracted by an emerging thought and, when distracted, the repeated mantra can be used to refocus the mind and release the distracting thought.

Final thoughts

ALTHOUGH THE INDUSTRIAL AND TECHNOLOGICAL EXPANSION SEEN IN THE WEST HAS UNDOUBTEDLY FACILITATED OUR ADVANCEMENT AS A SPECIES, IT HAS ALSO REEKED HAVOC ON OUR ECOSYSTEM AND OUR SENSE OF WELLBEING, BOTH AS INDIVIDUALS AND AS A GLOBAL SOCIETY. The alarming rise of stress-related diseases in recent years, together with the emergence of a number of worrying health issues linked with our non-sustainable lifestyles, has placed an enormous burden on our Western healthcare systems. Our traditional Western approach to healthcare fails to offer a complete solution for these escalating healthcare issues. There has been, therefore, a shift towards a more holistic, patient-centred approach to healthcare that acknowledges both the pivotal role played by individuals in maintaining their health and managing disease and the potential role played by therapies, such as meditation, in assisting individuals in achieving these goals. The widespread acceptance of meditation as a possible therapeutic intervention, both to prevent disease and to manage disease when it occurs, has, until recently, been hindered by the lack of accessible information about meditation and its effects on measurable health outcomes. This book has attempted to address this need and, in doing so, provides the scientific rationale for the use of meditation in the clinical setting.

Our brains are undoubtedly a mind-boggling feat of natural engineering. In the 21st century, even with all of our technical wizardry, we are only beginning to grasp the complexity of the human brain and the behaviour it so efficiently controls and coordinates. The human brain is highly adaptable; networks of brain cells can be moulded into an infinite number of different configurations, and this innate plasticity makes each one of us unique. Like an imprint created by pressing a key into a piece of putty, our brain records an imprint of our experiences, and this determines our memory of the past and our behaviour in the future. The plasticity of our brains also confers the ability to learn an astonishing range of different cognitive skills and behaviours; as humans, we can analyse evidence and create hypotheses to explain our observations, we can learn to perform complex movements or behaviours by rote, and we can remember huge volumes of text word for word or capture a beautiful memory for life. Our brains appear to have been designed to allow us ultimate freedom of expression. They provide a medium on which to imprint our experiences and learned behaviours. However, if this imprint is not updated in line with new experiences, our behaviours can become set and rigid, just like a piece of putty will become hard if left untouched for some time. Operating on automatic pilot, we are not exploiting the freedom of expression afforded to us; we are robots, prisoners of our conditioning. Our brains must always therefore be challenged to redefine our behaviour in line with our experiences. Learning should not stop after school or university. It is an ongoing process through which it is possible to optimize the performance of our brain, tap into our expressive potential and reap the benefits of such a magnificent piece of neural engineering.

The complexity of the brain has forced a shift away from a reductionist approach to research towards a more integrated systems approach. This groundbreaking whole-brain research reveals similarities between the principles inherent in our brain's function and those evident in the teaching of the mystical traditions and the behaviour of matter at a quantum level. Although these similarities should not be interpreted as being more than anecdotally interesting, they do reveal an important strategy through which to further optimize the performance of our brains. When problem-solving, one often has to step back from the details of the problem to establish context and reveal possible solutions. This suggests that we have the innate ability to both zoom into the detail of an experience and zoom out to give the 'bigger picture'. In terms of brain function, this arises from our innate ability to switch between two different modes of thinking/perceiving; from left-brain thinking to right-brain thinking, and vice versa. Each hemisphere is credited with different attributes; no one attribute can be said to be universally 'better' or more desirable than another. By acknowledging these different modes, one can choose which mode of thinking/perceiving is the most appropriate at any one given time and set of circumstances. Our brains are therefore performing optimally within the restrictions placed by our existing neural circuitry.

At the same time, our brains also appear to mediate something a little less tangible. In addition to allowing us to perceive, interpret and remember a wide range of everyday experiences, such as the sight, smell and feel of an orange, our brains also mediate profound experiences, such as the mystical experiences elicited through meditation. Humans are not only hard-wired to experience ordinary reality, but they are also hard-wired to experience higher states of consciousness. A number of different types of meditation have been devised that unfold a chain of processes within the human brain that mediate our access to these higher states of consciousness. Recent brain imaging studies have revealed the brain regions involved in this chain of processes, and detailed brain wave recordings have suggested a correlation between different brain wave states and specific levels of consciousness. This research has prompted a detailed investigation of the neuropsychology of mystical experiences. It appears that our logical minds have evolved in an attempt to make sense of our everyday experiences; to seek out patterns and question the meaning of events and the purpose of our lives. The myths and rituals devised by our ancestors tap into the innate functioning of our brains and provide access to ways of knowing and thinking that often provide answers to these questions. The findings of these studies have partially demystified meditation, and they also provide clues about the role of meditation in optimizing the performance of our brains.

Our recent technological advances have also allowed us to

explore potential ways of facilitating meditation and our attainment of higher states of consciousness. Meditation requires many years of training before the individual can efficiently access higher states of consciousness. It is perhaps not surprising that, in our *'we want it and we want it now'* culture, considerable time and energy have been invested in the search for methods of fast-tracking this training process. Entheogenic substances can provide transient glimpses of altered or higher states of consciousness; however, they do not represent a viable long-term strategy through which to facilitate the attainment of these states. Biofeedback, on the other hand, has huge potential as a mystical technology. Regular use of the GSR can increase awareness of unconscious behaviour, and therefore offers a potential way of facilitating the relaxation response that underlies meditation. Regular neurofeedback can increase awareness of the internal states and their influence on behaviour. It can therefore be used to train the individual to switch between different internal states thereby matching their internal state to the task at hand. These mystical technologies therefore promise to support our attempts to meditate and optimize the performance of our brains, and therefore express more of our full potential as human beings.

Regular meditation produces measurable health benefits and elicits sustained improvement in the physical, psychological and emotional wellbeing of the practitioner. It offers benefits to the practitioner in terms of stress reduction, improved cardiovascular health and immune function, and improved coping strategies in the face of disease. It also elicits significant cognitive and psychological changes, and plays an important role in driving the personal development of the practitioner. It therefore has huge potential as a supportive therapy in the clinical setting. There is a strong body of evidence suggesting that meditation should in fact be seen as a crucial component of both preventative medicine and effective disease management. The integration of meditation with other more traditional methods of patient management is in line with sentiments of the current patient-focused healthcare environment, and the resulting holistic approach promises to deliver a safer, faster and more effective standard of patient care. The integration of meditation into our everyday lives is also warranted, both as a method of combating the stresses of modern life and preventing disease. Not only will this undoubtedly impact on the future demands placed on our healthcare resources, but it also empowers individuals to play a major role in maintaining their health and wellbeing.

However, for many, the physical, psychological and emotional benefits of meditation are, admittedly desirable, 'side effects' of regular practice. The true benefit of meditation is seen to stem from the connection, during meditation, with an extraordinary reality that transcends our everyday, limited, ego-driven reality. The self-transcendent nature of meditation and mystical experiences, that we have seen stems from the shift from left- to right-brain activity during meditation, brings a sense of interconnectedness, universal compassion and unbounded love. The spiritual growth evoked by these experiences can be remarkable if they are successfully integrated into the individual's everyday life. This element of meditation is something that will never be fully explained by science nor fully understood using rational thought. There comes a point, in any investigation, when it becomes necessary to experience something yourself to fully understand it. Meditation, and mystical experiences, appear to be part and parcel of what it means to be human; our brains are hard-wired to access higher levels of consciousness. All we need to do is give it a go.

References

Chapter 1: Meditation: what is it and why do we need it?

1 Kabat-Zinn, J., 'Mindfulness-based interventions in context: past, present, and future', *Clinical Psychology: Science and Practice*, 2003: 10:144–56.

2 Saper, R.B., Eisenberg, D.M., Davis, R.B., Culpepper, L. and Phillips, R.S., 'Prevalence and patterns of adult yoga use in the United States: results of a national survey'. *Alternative Therapies in Health and Medicine*, 2004:10:44–49.

3 Cohen, S., Hamrick, N., Rodriguez, M.S., Feldman, P.J., Rabin, B.S. and Manuck, S.B., 'Reactivity and vulnerability to stress-associated risk for upper respiratory illness', *Psychosomatic Medicine*, 2002: 64(2):302–10.

4 de Vernejoul, P. *et al*, 'Etudes des méridiens d'acupuncture par les traceurs radioactifs', *Bulletin de l'Académie Nationale de Médecine*, 1991: 169:1071–5.

Chapter 2: Peering beneath the skull: how the brain works

1 Nussbaum, M.C., *Aristotle: Aristotle's de Motu Animalium*, Princeton University Press, 1990.

2 Asimov, I., in foreword to Hooper, J. and Teresi, D., *The Three-Pound Universe*, Jeremy P. Tarcher, 1986.

3 Hebb, D.O., *The Organization of Behaviour*, John Wiley and Sons, 1949.

4 Peat, D., *Infinite Potential: the life and times of David Bohm*, Addison-Wesley, 1997.

Chapter 3: Meditation and mystical experiences

1 D'Aquili, E. and Newberg, A.B., *The Mystical Mind: probing the biology of religious experience*, Augsburg Fortress Publishers, 1999.

2 Salmon, P., Septon, S., Weissbecker, I., Hoover, K., Ulmer, C. and Studts, J.L., 'Mindfulness meditation in clinical practice', *Cognitive and Behavioural Practice*, 2004: 11:434–46.

3 Aftanas, L.I. and Golocheikine, S.A., 'Non-linear dynamic complexity of the human EEG during meditation', *Neuroscience Letters*, 2002: 330(2):143–6.

4 Hebert, R. and Lehmann, D., 'Theta bursts: an EEG pattern in normal subjects practising the transcendental meditation technique', *Electroencephalography and Clinical Neurophysiology*, 1977: 42(3):397–405.

5 Lutz, A., Greischar, L.L., Rawlings, N.B., Ricard, M. and Davidson, R.J., 'Long-term meditators self-induce high-amplitude gamma synchrony during mental practice', *Proceedings of the National Academy of Sciences*, USA, 2004: 101(46):16369–73.

6 Aftanas, L.I. and Golocheikine, S.A., 'Human anterior and frontal midline theta and lower alpha reflect emotionally positive state and internalized attention: high-resolution EEG investigation of meditation', *Neuroscience Letters*, 2001: 310(1):57–60.

7 Cade, M. and Coxhead, N., *The Awakened Mind: biofeedback and the development of higher states of awareness*, Delacorte Press/Eleanor Friede, 1979.

Chapter 4: Bridging science and spirituality

1 Larson, E.J. and Witham, L., 'Scientists and religion in America', *Scientific American*, 1999.

2 Einstein, A., *Ideas and Opinions*, Three Rivers Press, 1982.

3 Wittgenstein, L., *Logico-Philosophicus*, translated by Pears, D.F. and McGuinness, B.F., Routledge and Kegan Paul, 1961.

4 Blackmore, S.J., 'Unrepeatability: parapsychology's only finding', in: Shapin, B. and Coly, L. (eds), *Repeatability Problem in Parapsychology*, Parapsychology Foundation, 1985: 183–206; Fromm, E., *Psychoanalysis and Religion*, Yale University Press, 1950.

5 Vyse, S.A., *Believing in Magic: the psychology of superstition*, Oxford University Press (NY), 1997.

6 Campbell, J., *The Hero with a Thousand Faces*, Princeton University Press, 1968 (2nd ed.).

7 Jung, C.G., *Recent Thoughts on Schizophrenia*, Routledge & Kegan Paul, 1957.

8 Horgan, J., *Rational Mysticism: dispatches from the border between science and spirituality*, Houghton Mifflin Books, 2003.

Chapter 5: Hard-wired: a high-performance mind

1 Doblin, R., 'Pahnke's "Good Friday Experiment": a long-term follow-up and methodological critique', *Journal of Transpersonal Psychology*, 1991: 23(1).

2 Abraham, H.D., Aldridge, A.M. and Gogia, P., 'The psychopharmacology of hallucinogens', *Neuropsychopharmacology*, 1996: 14(4):285–98.

3 Cohen, S., *The Beyond Within: the LSD story*, Athenaeum Press, 1964 (cited in: Novak, S.J., 'Second thoughts on psychedelic drugs', *Endeavour*, 1998: 22(1):21–3).

4 Krupitsky, E.M. and Grinenko, A.Y., 'Ketamine psychedelic therapy (KPT): a review of the results of ten years of research', *Journal of Psychoactive Drugs*, 1997: 29(2):165–83.

5 Evans, C.O. and Fudjack, J., *Consciousness: an interdisplinary study*, 1976. Available online at www.mentalstates.net/consciousness.html.

6 Vernon, D., Egner, T., Cooper, N., Compton, T., Neilands, C., Sheri, A. and Gruzelier, J., 'The effect of training distinct neurofeedback protocols on aspects of cognitive performance', *International Journal of Psychophysiology*, 2003: 47(1):75–85.

7 Tart, C., *Altered States of Consciousness*, HarperCollins, reprinted 1990.

Chapter 6: Meditation and health

1 Roth, B. and Robbins, D., 'Mindfulness-based stress reduction and health-related quality of life: findings from a

bilingual inner-city patient population', *Psychosomatic Medicine*, 2004: 66(1):113–23.

2 Walton, K.G., Fields, J.Z., Levitsky, D.K., Harris, D.A., Pugh, N.D. and Schneider, R.H., 'Lowering cortisol and CVD risk in postmenopausal women: a pilot study using the transcendental meditation program', *Annals of the New York Academy of Sciences*, 2004: 1032:211–5.

3 Infante, J.R., Torres-Avisbal, M., Pinel, P., Vallejo, J.A., Perán, F., Gonzalez, F., Contreras, P., Pacheco, C., Roldan, A. and Latre, J.M., 'Catecholamine levels in practitioners of the transcendental meditation technique', *Physiology and Behavior*, 2001: 72(1–2):141–6.

4 Grossman, P., Niemann, L., Schmidt, S. and Walach, H., 'Mindfulness-based stress reduction and health benefits: a meta-analysis,' *Journal of Psychosomatic Research*, 2004: 57(1):35–43.

5 Shapiro, S.L., Schwartz, G.E. and Bonner, G., 'Effects of mindfulness-based stress reduction on medical and premedical students', *Journal of Behavioral Medicine*, 1998: 21(6):581–99.

6 Rosenzweig, S., Reibel, D.K., Greeson, J.M., Brainard, G.C. and Hojat, M., 'Mindfulness-based stress reduction lowers psychological distress in medical students', *Teaching and Learning in Medicine*, 2003: 15(2):88–92.

7 Roth, B. and Stanley, T.W., 'Mindfulness-based stress reduction and healthcare utilization in the inner city: preliminary findings', *Alternative Therapies in Health and Medicine*, 2002: 8(1):60–2, 64–6.

8 Kabat-Zinn, J., Wheeler, E., Light, T., Skillings, A., Scharf, M.J., Cropley, T.G., Hosmer, D. and Bernhard, J.D., 'Influence of a mindfulness meditation-based stress reduction intervention on rates of skin clearing in patients with moderate to severe psoriasis undergoing phototherapy (UVB) and photochemotherapy (PUVA)', *Psychosomatic Medicine*, 1998: 60(5):625–32.

9 Wilson, A.F., Honsberger, R., Chiu, J.T. and Novey, H.S.,

REFERENCES

'Transcendental meditation and asthma', *Respiration*, 1975: 32(1):74–80.

10 Keefer, L. and Blanchard, E.B., 'The effects of relaxation response meditation on the symptoms of irritable bowel syndrome: results of a controlled treatment study', *Behaviour Research and Therapy*, July 2001: 39(7):801–11.

11 Sudheesh, N.N. and Joseph, K.P., 'Investigation into the effects of music and meditation on galvanic skin response', *ITBM-RBM*, 2000: 21(3):158–63.

12 Muskatel, N., Woolfolk, R.L., Carrington, P., Lehrer, P.M. and McCann, B.S., 'Effect of meditation training on aspects of coronary-prone behavior', *Perceptual and Motor Skills*, 1984: 58(2):515–18.

13 Davidson, R.J., Kabat-Zinn, J., Schumacher, J., Rosenkranz, M., Muller, D., Santorelli, S.F., Urbanowski, F., Harrington, A., Bonus, K. and Sheridan, J.F., 'Alterations in brain and immune function produced by mindfulness meditation', *Psychosomatic Medicine*, 2003: 65(4):564–70.

14 Solberg, E.E., Halvorsen, R., Sundgot-Borgen, J., Ingjer, F. and Holen, A., 'Meditation: a modulator of the immune response to physical stress? A brief report', *British Journal of Sports Medicine*, 1995: 29(4):255–7.

15 Antoni, M.H., Cruess, D.G., Klimas, N., Carrico, A.W., Maher, K., Cruess, S., Lechner, S.C., Kumar, M., Lutgendorf, S., Ironson, G., Fletcher, M.A. and Schneiderman, N., 'Increases in a marker of immune system reconstitution are predated by decreases in 24-h urinary cortisol output and depressed mood during a 10-week stress management intervention in symptomatic HIV-infected men', *Journal of Psychosomatic Research*, 2005: 58(1):3–13.

16 Sanchez-Barcelo, E.J., Cos, S., Fernandez, R. and Mediavilla, M.D., 'Melatonin and mammary cancer: a short review', *Endocrine-related Cancer*, 2003: 10(2):153–9.

17 Lissoni, P., Malugani, F., Brivio, F., Piazza, A., Vintimilla, C., Giani, L. and Tancini, G., 'Total pineal endocrine

substitution therapy (TPEST) as a new neuroendocrine palliative treatment of untreatable metastatic solid tumor patients: a phase II study', *Neuroendocrinology Letters*, 2003: 24(3–4):259–62.

18 Tooley, G.A., Armstrong, S.M., Norman, T.R. and Sali, A., 'Acute increases in night-time plasma melatonin levels following a period of meditation', *Biological Psychology*, 2000: 53(1):69–78.

19 Massion, A.O., Teas, J., Hebert, J.R., Wertheimer, M.D. and Kabat-Zinn, J., 'Meditation, melatonin and breast/prostate cancer: hypothesis and preliminary data', *Medical Hypotheses*, 1995;44(1):39–46.

20 Carlson, L.E., Speca, M., Patel, K.D. and Goodey, E., 'Mindfulness-based stress reduction in relation to quality of life, mood, symptoms of stress and levels of cortisol, dehydroepiandrosterone sulfate (DHEAS) and melatonin in breast and prostate cancer outpatients', *Psychoneuro-endocrinology*, 2004: 29(4):448–74.

21 Speca, M., Carlson, L.E., Goodey, E. and Angen, M., 'A randomized, wait-list controlled clinical trial: the effect of a mindfulness meditation-based stress reduction program on mood and symptoms of stress in cancer outpatients', *Psychosomatic Medicine*, 2000: 62(5):613–22.

22 Solberg, E.E., Ekeberg, O., Holen, A., Ingjer, F., Sandvik, L., Standal, P.A. and Vikman, A., 'Hemodynamic changes during long meditation', *Applied Psychophysiology and Biofeedback*, 2004: 29(3):213–21.

23 Peng, C.K., Henry, I.C., Mietus, J.E., Hausdorff, J.M., Khalsa, G., Benson, H. and Goldberger, A.L., 'Heart rate dynamics during three forms of meditation', *International Journal of Cardiology*, 2004: 95(1):19–27.

24 Telles, S., Joshi, M., Dash, M., Raghuraj, P., Naveen, K.V. and Nagendra, H.R., 'An evaluation of the ability to voluntarily reduce the heart rate after a month of yoga practice', *Integrative Physiological and Behavioral Science*, 2004: 39(2):119–25.

25 Manikonda, P., Stoerk, S., Toegel, S., *et al.*, 'Influence of non-pharmacological treatment (contemplative meditation and breathing technique) on stress induced hypertension: a randomised controlled study', *American Journal of Hematology*, 2005: 18(5 pt 2):89–90A; abstract P232.

26 Barnes, V.A., Davis, H.C., Murzynowski, J.B. and Treiber, F.A., 'Impact of meditation on resting and ambulatory blood pressure and heart rate in youth', *Psychosomatic Medicine*, 2004: 66(6):909–14.

27 Canter, P.H. and Ernst, E., 'The cumulative effects of transcendental meditation on cognitive function – a systematic review of randomised controlled trials', *Wien Klin Wochenschr.* 2003: 115(21–22):758–66.

28 Kutz, I., Leserman, J., Dorrington, C., Morrison, C.H., Borysenko, J.Z. and Benson, H., 'Meditation as an adjunct to psychotherapy: an outcome study', *Psychotherapy and Psychosomatics*, 1985: 43(4):209–18.

29 Delmonte, M.M., 'Meditation and anxiety reduction: a literature review', *Clinical Psychology Review*, 1985: 5:91–102.

30 Dillbeck, M.C., 'The effect of the transcendental meditation technique on anxiety level', *Journal of Clinical Psychology*, 1977: 33(4):1076–8.

31 Miller, J.J., Fletcher, K. and Kabat-Zinn, J., 'Three-year follow-up and clinical implications of a mindfulness meditation-based stress reduction intervention in the treatment of anxiety disorders', *General Hospital Psychiatry*, 1995: 17(3):192–200.

32 West, M.A., 'Meditation, personality, and arousal', *Personalty and Individual Differences*, 1980: 1:135–42.

33 Williams, P., Francis, A. and Durham, R., 'Personality and meditation', *Perceptual and Motor Skills*, 1976: 43(3 pt. 1):787–92.

34 Krisanaprakornkit, T., Krisanaprakornkit, W., Piyavhatkul, N. and Laopaiboon, M., 'Meditation therapy for anxiety disorders', *The Cochrane Database of Systematic Reviews*, 2007,

Issue 1.

35 Finucane, A. and Mercer, S.W., 'An exploratory mixed methods study of the acceptability and effectiveness of mindfulness-based cognitive therapy for patients with active depression and anxiety in primary care', *BMC Psychiatry*, 2006: 6:14.

36 Teasdale, J.D., Segal, Z.V., Williams, J.M., Ridgeway, V.A., Soulsby, J.M. and Lau, M.A., 'Prevention of relapse/recurrence in major depression by mindfulness-based cognitive therapy', *Journal of Consulting and Clinical Psychology*, 2000: 68(4):615–23.

37 Aron, A. and Aron, E.N., 'The transcendental meditation program's effect on addictive behavior', *Addict Behaviors*, 1980: 5(1):3–12.

38 Monahan, R.J., 'Secondary prevention of drug dependence through the transcendental meditation program in metropolitan Philadelphia', *International Journal of Addiction*, 1977: 12(6):729–54.

39 Davis, J.M., Fleming, M.F., Bonus, K.A. and Baker, T.B, 'A pilot study on mindfulness based stress reduction for smokers', *BMC Complementary and Alternative Medicine*, 2007: 7:2.

40 Kabat-Zinn, J., 'An outpatient program in behavioral medicine for chronic pain patients based on the practice of mindfulness meditation: theoretical considerations and preliminary results', *General Hospital Psychiatry*, 1982: 4(1):33–47.

41 Kabat-Zinn, J., Lipworth, L. and Burney, R., 'The clinical use of mindfulness meditation for the self-regulation of chronic pain', *Journal of Behavioral Medicine,* 1985: 8(2):163–90.

42 Horton-Deutsch, S., O'Haver Day, P., Haight, R. and Babin-Nelson, M., 'Enhancing mental health services to bone marrow transplant recipients through a mindfulness-based therapeutic intervention', *Complementary Therapies in Clinical Practice,* 2007, doi 10.1016/j.ctcp.2006.11.0003.

43 Shapiro, D.H. Jr, 'Adverse effects of meditation: a

preliminary investigation of long-term meditators', *International Journal of Psychosomatic Medicine,* 1992: 39(1–4):62–7.

44 West, M.A., 'Meditation, personality, and arousal', *Personalty and Individual Differences,* 1980: 1:135–42.

Chapter 7: Meditation in our everyday lives

1 Kabat-Zinn, J., *Full Catastrophe Living: how to cope with stress, pain and illness using mindfulness meditation,* 15th anniversary edition, Piatkus Books, 2006.

2 Nataraja, K., *Dancing With Your Shadow: integrating the ego and the self on the spiritual path,* Medio Media, 2006.

Further reading

Austin, J.H., *Zen and the Brain: toward an understanding of meditation and consciousness,* MIT Press, reprint edition 1999.

Cade, M. and Coxhead, N., *The Awakened Mind: biofeedback and the development of higher states of awareness,* Delacorte Press/Eleanor Friede, 1979.

Canter, P.H., 'The therapeutic effects of meditation', *BMJ,* 2003: 326(7398):1049–50.

Capra, F., *The Tao of Physics: an exploration of the parallels between modern physics and eastern mysticism,* (anniversary edition), Shambhala Publications, Inc., 2000.

D'Aquili, E. and Newberg, A.B., *The Mystical Mind: probing the biology of religious experience,* Augsburg Fortress Publishers, 1999.

Horgan, J., *Rational Mysticism: dispatches from the border between science and spirituality,* Houghton Mifflin Books, 2003.

Huxley, A., *Doors of Perception,* Chatto and Windus, 1954.

Huxley, A., *The Perennial Philosophy,* Harper and Row, 1945.

Kabat-Zinn, J., 'Mindfulness-based interventions in context: past, present, and future', *Clinical Psychology: Science and Practice,* 2003: 10:144–56.

Kit, W.K., *The Complete Book of Tai Chu Chuan: a comprehensive guide to the principles and practice,* Element Books, 1996.

Newberg, A., d'Aquili, E. and Rause, V., *Brain Science and the Biology of Belief: why God won't go away,* Ballantine Books, 2001.

Reid, D., *Chi-Gung: harnessing the power of the universe,* Simon and Schuster, 1998.

Robbins, J., *A Symphony in the Brain: the evolution of the new brain wave biofeedback,* Grove Press, 2001.

Satinova, J., *The Quantum Brain: the search for freedom and the next generation of man,* John Wiley and Sons, 2001.

Vyse, S.A., *Believing in Magic: the psychology of superstition,*

Oxford University Press (NY), 1997.

Wilber, K., *No Boundary: eastern and western approaches to personal growth*, Shambhala Publications, 1979, 2001.

Wise, A., *The High-Performance Mind: mastering brainwaves for insight, healing, and creativity*, Jeremy P. Tarcher, 1997.

Index

INDEX

Page numbers in *italic* refer to the illustrations

A

abstract concepts 57

abstractive operator 121, 122, 123

acceptance, mindfulness-based stress reduction (MBSR) and 197

active meditation 84, *91*, 93

acupuncture
flow of ch'i 38, 41
meridians 22
traditional Chinese medicine 39

addictive behaviour, benefits of meditation 180–1

ADHD (attention deficit hyperactivity disorder) 12–13, 162

adrenal glands 60

adrenaline 171

affective disorders 162

Aftanas 101

AIDS 37

alcohol 188
alcoholism 148, 149
meditation to reduce intake of 176, 180
serotonin production and 34
stress and consumption of 28, 30

alpha waves 97
and levels of consciousness 104–5, 106, 107
in meditation 99, 101
neurofeedback 163, 164, 165, 167

American Medical Association (AMA) 148

amygdala 58, 59
cognitive operators and 123
functions 61–2
meditation and 87–9, 92

Anderson (MD) Wellness Programme 189

anger, side-effects of meditation 183

anima, archetype 133

animal movements, chi kung 22

animus, archetype 133

anterior cingulate cortex (ACC) 158–60, *159*

anxiety
benefits of meditation 171, 179, 180, 182
and healthcare costs 37
neurofeedback and 162
psychological impact of illness 31
stress and 28, 32, 34

archetypes 119–20, 129, 133, 136

Aristotle 45

arousal
anterior cingulate cortex 159
biofeedback and 156
controlling levels of 160–1
fear of potential threats 123–4
'fight or flight' reaction 32, 60
galvanic skin response (GSR) 13, 156–8, 160–2
in meditation 162
thalamus and 59

asana, yoga 21

Asimov, Isaac 49

association areas, brain *80*
functions 54, 56, 57, 58
in mystical experiences 81–3

associative memory 66–7, *66*

astanga yoga 21

asthma 29, 172

astronomy 110

atherosclerosis 176

attention 19, 95–6
active meditation 90–1
benefits of meditation 177
centering prayer 25–6
neurofeedback and 166
in tai chi 24

attention association area *80*
functions 56, 82–3
in meditation 84, 91
and mystical experiences 81

attention deficit hyperactivity disorder (ADHD) 12–13, 162

authoritarian religions 128–9

autism 123

automatic behaviour 194

INDEX

Acknowledgements

Executive Editor Sandra Rigby
Managing Editor Clare Churly
Executive Art Editor Leigh Jones
Illustrators BrindeauMexter
Page make-up Dorchester
  Typesetting Group Ltd
Senior Production Controller
  Simone Nauerth